Solid Mechanics and Its Applications

Volume 232

The fundamental questions arising in mechanics are: Why?, How?, and How much? The aim of this series is to provide lucid accounts written by authoritative researchers giving vision and insight in answering these questions on the subject of mechanics as it relates to solids. The scope of the series covers the entire spectrum of solid mechanics. Thus it includes the foundation of mechanics; variational formulations; computational mechanics; statics, kinematics and dynamics of rigid and elastic bodies; vibrations of solids and structures; dynamical systems and chaos; the theories of elasticity, plasticity and viscoelasticity; composite materials; rods, beams, shells and membranes; structural control and stability; soils, rocks and geomechanics; fracture; tribology; experimental mechanics; biomechanics and machine design. The median level of presentation is the first year graduate student. Some texts are monographs defining the current state of the field; others are accessible to final year undergraduates; but essentially the emphasis is on readability and clarity.

Springer and Professors Barber and Klarbring welcome book ideas from authors. Potential authors who wish to submit a book proposal should contact Dr. Mayra Castro, Senior Editor, Springer Heidelberg, Germany, email: mayra.castro@springer.com

Indexed by SCOPUS, Ei Compendex, EBSCO Discovery Service, OCLC, ProQuest Summon, Google Scholar and SpringerLink.

Arul Jayachandran

Elastic Postbuckling Analysis of Structural Systems

Computational Methods and Applications

Springer

Arul Jayachandran
Structural Engineering Laboratory
Department of Civil Engineering
Indian Institute of Technology Madras
Chennai, Tamil Nadu, India

ISSN 0925-0042 ISSN 2214-7764 (electronic)
Solid Mechanics and Its Applications
ISBN 978-3-031-89935-5 ISBN 978-3-031-89936-2 (eBook)
https://doi.org/10.1007/978-3-031-89936-2

This Springer imprint is published by the registered company Springer Nature Switzerland AG
The registered company address is: Gewerbestrasse 11, 6330 Cham, Switzerland

This book is dedicated to the Greater Glory of the Lord.

Foreword

Through my many years on the Structural Stability Research Council, I had the absolute pleasure of meeting and learning about the impressive research conducted by Prof. Arul Jayachandran. I also share an academic lineage with several esteemed Cornell University professors, now late, who were pioneers in structural stability, including George Winter, William McGuire, and Richard Gallagher. Their contributions, alongside the classic works of Bleich, Timoshenko, Vlasov, and Galambos, have provided our profession with the fundamental knowledge to compute the buckling strengths of structural cross-sections, members, and systems.

However, during my work on SSRC's *Guide to Stability Design Criteria for Metal Structures*, I came to fully appreciate the significance of understanding postbuckling behavior. Despite a well-established body of research in this area, few, if any, books have focused exclusively on this subject—until now.

Accurately predicting and analyzing postbuckling responses are essential for both fundamental research and practical engineering applications. *Elastic Postbuckling Analysis of Structural Systems: Computational Methods and Applications* by Prof. Arul Jayachandran offers a comprehensive and methodical treatment of this critical topic, mainly through computational approaches.

The book systematically develops the secant matrix approach within the total potential energy framework, providing a structured methodology for postbuckling analysis. A key contribution is the introduction of the Variable Order Secant Matrix (VOSM) method, which refines the handling of geometric nonlinearities in various structural systems, including beams, frames, and thin-walled sections. By integrating theoretical foundations with computational implementations, this text is a good resource for researchers seeking to develop or improve numerical models in structural stability.

A notable strength of this work is its emphasis on problem formulation and numerical validation. Through benchmark problems and case studies, the book ensures that the methodologies discussed are both applicable and verifiable. The discussions on nonlinear finite element formulations, stability indicators, and path-tracing techniques provide a clear and practical framework for solving complex postbuckling problems.

Professor Jayachandran's work is a well-organized and significant contribution to computational structural stability. We are fortunate to benefit from an author who seamlessly blends his academic expertise with extensive engineering practice. I am confident that this book will serve as an excellent resource for graduate students, researchers, and engineers seeking to develop computational tools for structural stability investigations.

This book earns its place among the classics on my bookshelf, and I trust it will find a valued place on yours as well—serving as an outstanding reference for those engaged in geometric nonlinear stability analysis and finite element modeling.

February 2025

Ronald D. Ziemian
Honorary Editor,
Journal of Constructional Steel Research, Elsevier
Bucknell University
Lewisburg, USA

Preface

This book focuses on computational methods for structural stability, with an emphasis on the total potential energy (TPE) framework. TPE is an effective tool for analyzing the stability of various structural systems, including reticulated, framed, and continuum structures. My early work in structural stability relied on the strength of the TPE framework, and this book extends that foundation by incorporating the secant matrix technique, also known as the 'N1-N2 method' by Rajasekaran and Murray. This method allows for multiple formulations under the TPE framework.

The book presents geometric nonlinear finite element formulations using the secant matrix technique for elastic postbuckling analysis. These formulations are applied to various structural systems, including beams, frames, arches, thin-walled sections, trusses, shells, and plates—perfect and imperfect. The geometrically nonlinear formulations are developed using the Total Lagrangian (TL) and co-rotational TL formats to address large displacements and rotations encountered in postbuckling states.

For structural systems like beams, frames, arches, and thin-walled open sections, conventional methods often face challenges due to large deformations. A new Variable Order Secant Matrix (VOSM) method introduced in this book offers an accurate postbuckling analysis by employing higher-order strain-displacement relations. This VOSM approach automatically limits the order of the secant matrices based on the nonlinearity encountered in the load-deformation behavior in the postbuckled state. The VOSM method is extended to analyze frames and thin-walled open sections, while Marguerre's shell theory is employed to model the postbuckling behavior of shallow shells.

These nonlinear finite element formulations are implemented in a custom computer program incorporating the minimum residual displacement (MRD) method developed by Prof. Chan S. L. The 'current stiffness parameter' is used as a stability indicator, guiding the postbuckling path-tracing algorithm.

In addition to serving as a repository of early contributions to the use of finite elements in structural stability investigations, this book presents a large number of benchmark problems that are crucial for validating postbuckling analysis methods. These benchmark problems provide a valuable reference for researchers and practitioners.

Thematically, this book: (i) documents research on the application of finite elements in postbuckling stability investigations; (ii) establishes a TPE-based secant matrix computational framework for postbuckling analysis of structural systems; and (iii) provides benchmark problems to support further research and practical applications.

Chennai, India

Arul Jayachandran

Acknowledgments

'Rejoice always, pray continually, give thanks in all circumstances; for this is God's will for you in Christ Jesus'
—1 Thessalonians 5:16–18.

Thank you, Lord Jesus, in all humility. Your great Glory and Grace have sustained me through every valley I have walked.

I owe much to the Indian Institute of Technology Madras (IITM), Chennai, where my passion for structural stability first took root. I was fortunate to have my Master's and Ph.D. guided by professors who were alumni of Cornell University, USA—a beacon of excellence in structural stability. The introductory classes on structural stability by Prof. L. N. Ramamurthy were nothing short of a theatrical art form, leaving a profound mark on my journey. I owe my research in structural stability to Prof. V. Kalyanaraman, a genius, and a personal role model. To both my Gurus, Pranams.

My time as a visiting Associate Professor at Georgia Tech, Atlanta, furthered my structural stability research when Prof. Don White introduced me to the direct analysis method. The annual Structural Stability Research Council meetings in the USA continue to fuel my passion for structural stability.

I would like to express my gratitude to Prof. S. Rajasekaran, an accomplished emeritus professor at PSG Tech, Coimbatore, whose phenomenal work and insights into the secant matrix technique inspired my research in this area.

The Structural Engineering Research Centre (SERC) in Chennai was my home for 22 years, where I served as a Scientist in the Steel Structures Lab. I am indebted to SERC and my colleagues for the experiences, collaborations, and opportunities shaping my career.

To my students at IIT Madras, you are my greatest strength. Our beloved Room #402 at the Structural Engineering Lab has been a springboard for many bright futures. You 402ans offered me crucial assistance as I drafted this book. I am also thankful for the support of my faculty colleagues from the Structural Engineering Lab at IIT Madras.

Becoming an author was once a distant dream, made possible only by the unyielding love, patience, and support of my wife, Er. P. Premalatha. Thank you, my beloved, for your fervent prayers and unshakable belief in my work throughout these years. To my dear daughter, Ms. A. Monica, my son-in-law, Mr. K. Vinoth, and little Jordan-your encouragement has been my source of strength. I am also grateful to all my relatives from both sides of the family for their ever-present care and support.

Lastly, to my dear parents—Mom Jeyaseeli and Dad Sanjeevi—there are no words to express my gratitude for your sacrifices. Despite the hardships, you provided me with an education; I am eternally grateful for that. Praise be to God for his blessings throughout my journey.

Arul Jayachandran

Contents

Abbreviations

ADF	Additional displacement formulation
BFS	Bogner, Fox, Schmidt
CCF	Core congruential formulation
CLPT	Classical laminated plate theory
CR	Co-rotational
CR-TL	Co-rotational Total Lagrangian
CR-UL	Co-rotational Updated Lagrangian
CSP	Current stiffness parameter
DOF	Degree of freedom
EB	Euler–Bernoulli
FOST	First-order shear deformation theory
MRD	Minimum residual displacement
TDF	Total displacement formulation
TL	Total Lagrangian
TPE	Total potential energy
UL	Updated Lagrangian
VOSM	Variable Order Secant Matrices

Chapter 1
Introduction to Elastic Postbuckling Behavior of Structures

Abstract This chapter deals with the elastic postbuckling behavior of structural systems, beginning with an overview of structural stability fundamentals and later the critical points, bifurcation, and limit points. Using the total potential energy (TPE) framework, various postbuckling paths encountered in stability analysis are discussed. A review of the secant matrix technique and its applications across different structural types, such as trusses, beams, and thin-walled sections, are presented. The chapter also covers computational methods, focusing on iterative and incremental approaches, particularly the arc-length method for path following in postbuckling analysis. In summary, the computational and analytical bases are presented for a deep understanding of the postbuckling response of structures in practical applications.

Keywords Postbuckling analysis · Secant matrix · Critical points in structural stability · Bifurcation · Nonlinear structural stability

1.1 General

The evolution of the structural stability theory has been remarkable from the years of classical works in the eighteenth century by Euler and Lagrange and in the nineteenth century by Poincare and Lyapunov. Further advances took place in the latter part of the twentieth century through the work of several researchers like Koiter, Budiansky, Hutchinson, Zeigler, Timoshenko, Britvec, Chilver, Roorda, Thomson, Hunt, Supple, and several others. The classical work of Koiter (1945) was path-breaking as it laid the analytical framework for solutions to many problems in structural stability, consistent with experimental results. Theoretical knowledge and experimental investigations on the static response of nonlinear elastic structures have progressed considerably over the last forty years, and the understanding of elastic stability can now be considered good. We also see that the research in structural stability started focusing more on computational structural stability (Arboz and Hol 1990; Casciaro et al. 1992; Flores and Godoy 1992; Godoy et al. 1995; Santos 2012; Mejia-Nava et al. 2022) than on the analytical stability theory. Later, computational methods became very useful in investigating the structures beyond the linear critical loads. Thus, the

A. Jayachandran, *Elastic Postbuckling Analysis of Structural Systems*, Solid Mechanics and Its Applications 232, https://doi.org/10.1007/978-3-031-89936-2_1

research on tracing the stability behavior of structures in the postbuckling state gained momentum. This book focuses on the computational methods for investigating elastic structural systems in their postbuckled state.

1.2 Elastic Stability of Structural Systems

Investigations on computational and analytical structural stability, reported in the literature, have generally assumed the linear elastic material behavior. Such an assumption may look trivial and unimportant because most practical buckling behaviors are not completely elastic. However, instability is an essential property of structures at their extremes of geometry, like long slender struts, thin flat plates, and thin cylindrical shells, wherein the onset of instability frequently occurs while the material is still elastic. The instability of structures occurs due to changes in their physical state. Elastic instability is very common in many practical, light metal structures. However, the load-deformation relationship may be nonlinear due to a large deflection, better known as 'geometric nonlinearity'. Thus, the characteristic of instability is that large distortion may occur in the structure due to small changes in the external loads. At the same time, the material still behaves according to Hooke's law! Systems commonly exhibiting this property are two- and three-dimensional light structures such as light trusses, frameworks, thin-plated elements, reticulated shells, and geodesic domes. For several structural systems, the onset of motion or instability is not the end of the road in terms of load-carrying capacity. Structural systems may also carry a load beyond the initiation of buckling by mobilizing some reserve strength known as the 'postbuckling reserve strength'. Thin plates exhibit this phenomenon.

For a safe design of such instability-prone structural systems, it is necessary to detect the unstable equilibrium states to prevent failure due to geometry change. The designer, therefore, is faced with the problem of establishing the actual critical state of the system to ascertain if collapse by motion is imminent in such equilibrium states. Necessarily, the finite deformation of the whole structure and, hence, the geometric nonlinearity must be considered in such an analysis. Another problem encountered in such systems is that several structures are very sensitive to minute geometrical changes and also sensitive to small initial geometric imperfections. These can radically alter a structure's equilibrium configurations, and such considerations are practically significant in the design of slender space structures and light but highly stiff reticulated and continuous shells. The following sections explain the theory behind advanced stability investigations.

1.3 Critical Points Encountered in the Postbuckled State

Let us learn about the critical points of interest in structural stability and, later, the paths of equilibrium we encounter during stability investigations. Designers' desire to identify the critical equilibrium states reduces to finding the lowest critical point on the load-deformation path of a structure. As discussed earlier, the designers' interest also lies in what happens after traversing the critical points. This interest naturally comes, if the designer wants to take advantage of the postbuckled reserve strength in the structural system.

1.3.1 Limit and Bifurcation Points

Using the well-established concept of generalized coordinates in the theory of elastic stability, the deflected form of a structure (subjected to proportional loading $\lambda\mathbf{q}$, where $\mathbf{q}$ is the load vector corresponding to the generalized coordinates and λ is the proportionality constant) with a system of 'n' degrees of freedom (D.O.F.) would be defined entirely by a set of 'n' generalized displacement coordinates $\mathbf{p}_i$ ($i = 1,2,\ldots\ldots n$). A simple two D.O.F. representation of the Euler column using generalized coordinates p_1 and p_2 is shown in Fig. 1.1. In the first case (Fig. 1.1a), p_1 represents the axial shortening in the prebuckling state with stiffness EA/L and a reduced stiffness beyond λ_b. Displacement p_2 is active after the column buckling; hence, it is called 'the leading degree of freedom of bifurcation p_b'. Since the λ-p_b describes the instability phenomena, such degrees of freedom are usually of concern to the structural designers.

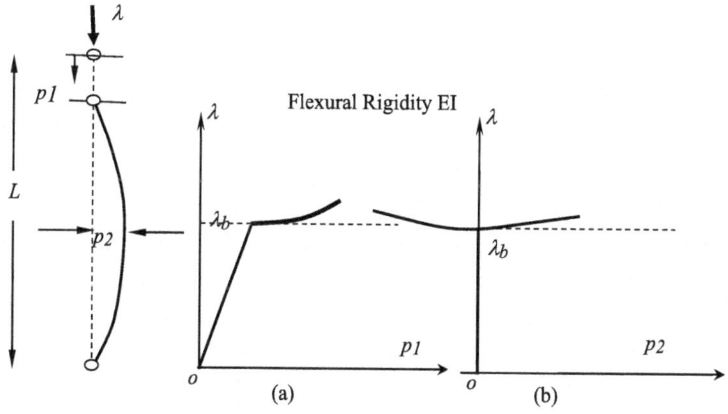

Fig. 1.1 Two degrees-of-freedom representation of an Euler column

In the case of a general 'n' D.O.F. system, the equilibrium path is often described by a projection of $\mathbf{p}(\lambda)$ on the plane $(\mathbf{p}_i, \lambda)$, where $\mathbf{p}_i$ is one of the generalized displacements of the structure, usually the leading degree of freedom. If the path passes through the origin of the coordinates $(\mathbf{p}_i, \lambda)$ (i.e.) $\mathbf{p}(0) = 0$, then the path is called the fundamental path. In other words, the term fundamental equilibrium path refers to the path of a structure in load-configuration space, which occurs from the onset of loading. In Fig. 1.1b, we see that the fundamental path (for load levels less than λ_b) is a straight line tracing the λ axis. At λ_b the structure would be in a state of neutral equilibrium and retain any small disturbance along p_2. Loads above λ_b along the λ axis, the path $p_2 = 0$ is unstable; hence, the structure changes or 'bifurcates' the equilibrium into the secondary postbuckling path transverse to the λ axis.

In a general context, along a fundamental path, two types of critical points may occur, namely (i) limit point (L) at which the maximum value of load parameter is reached and (ii) bifurcation point (B) at which the fundamental (prebuckling) and secondary (postbuckling) paths intersect. This is shown in Fig. 1.2. We see that the former is a point of local maximum of the fundamental equilibrium path in the load-configuration space, and the latter is the point of intersection of the fundamental equilibrium path with another distinct equilibrium path.

In the computational stability analysis of structural systems, limit points (L) can be determined using geometric nonlinear analysis, and the bifurcation points (B) are obtained by eigenvalue analysis.

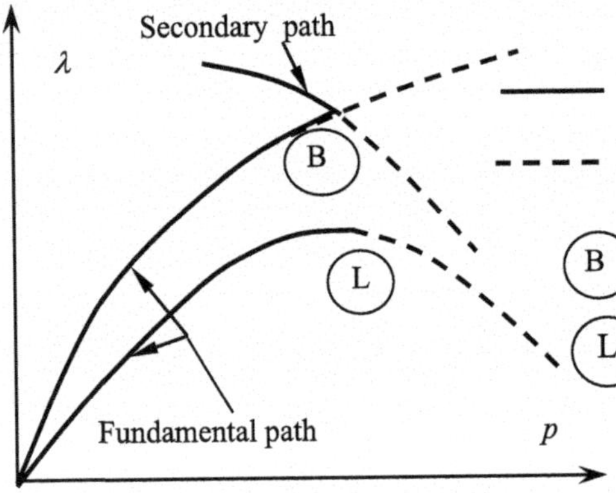

Fig. 1.2 Equilibrium paths and critical points

1.4 Various Postbuckling Paths Encountered in the Stability Investigations

Structural engineer's interest is often in the bifurcation point, because the structure loses stability at the lowest critical load. The bifurcation is considered 'distinct' when only one path emanates from the fundamental equilibrium path. The postbuckling equilibrium path, corresponding to a distinct point of bifurcation, may manifest itself in three different forms as (a) asymmetric, (b) stable symmetric, and (c) unstable symmetric postbuckling paths, as shown in Fig. 1.3. The 'initial imperfection' in geometry or loading of structures causes deformation of the leading degree of freedom from the onset of loading, and hence, the postbuckling paths invariably encounter limit points rather than bifurcation points, as shown in Fig. 1.3. The values of *a*, *b* in Fig. 1.3 are explained later.

In several structural systems with initial imperfections, the limit load may be reached at loads less than the critical value and hence causes a significant reduction in the load-carrying capacity. Structures that exhibit reduced load-carrying capacity due to initial imperfection are called 'imperfection-sensitive structures'. Loads above limit load are not tolerated in structural design. Nevertheless, recent codes of practice insist on the computation of the limit loads to compare with the service loads for understanding the safety against the limit point.

1.4.1 Identification of Postbuckling Paths Using Asymptotic Expansion of Total Potential Energy

In the subsequent sections, we see how these critical points are identified by (i) a potential energy-based analytical framework and (ii) a potential energy-based numerical framework.

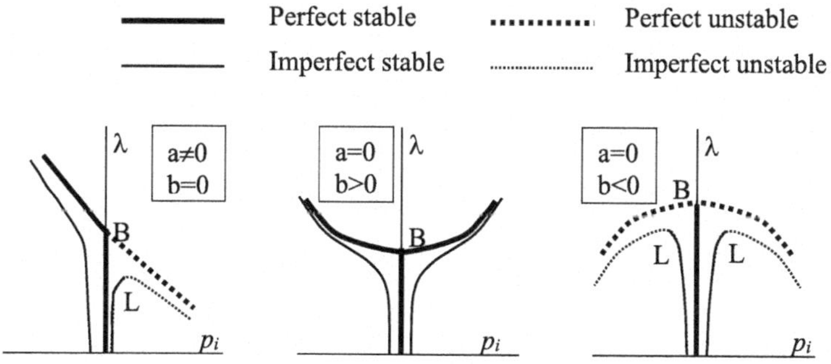

Fig. 1.3 Distinct postbuckling paths encountered in structural systems

The critical points and the structural behavior beyond the critical load can be described by analyzing the total potential energy (TPE) at various equilibrium states. This section briefly explains how critical points can be classified using the total potential energy.

In this book, Gibb's formalism represents vectors and matrices for rectangular Cartesian coordinate systems. Hence, a boldface letter type usually denotes a vector. The magnitude of the vector is denoted by the same letter in lightface type. The second-order (matrices) and higher-order tensors are represented by boldface capital letters (Malvern 1969; Flugge 1972).

Assuming that a structural system is subjected to conservative and proportional loads, the total potential energy 'π' of the system can be written as a function of the load parameter λ and the generalized displacement vector **p** as

$$\pi = \pi(\lambda, \mathbf{p}), \tag{1.1}$$

where $\pi(\lambda,\mathbf{p})$ is well-behaved single-valued function. Using Koiter's (1945) asymptotic approach to initial postbuckling behavior, the Taylor expansion of total potential energy can be written in the neighborhood of the critical load. This expansion can be carried out in terms of a small displacement path parameter 'ε' which may be the amplitude of the buckling mode. Thus, we may write

$$\mathbf{p} = \varepsilon \mathbf{p}_1 + \varepsilon^2 \mathbf{p}_2 + \varepsilon^3 \mathbf{p}_3 + \ldots, \tag{1.2}$$

and the load parameter λ as

$$\lambda = \lambda_{cr}(1 + a\varepsilon + b\varepsilon^2 + \ldots), \tag{1.3}$$

where λ_{cr} is the critical load factor. The total potential energy may be written as an expansion in functional form as

$$\pi(\lambda, \mathbf{p}) = \pi_2 + \pi_3 + \pi_4 + \ldots, \tag{1.4}$$

where the subscripts represent the order of the potential energy terms. The parameters a, b in Eq. (1.3) can be determined by the condition that

$$\frac{d\pi(\lambda, p)}{d\varepsilon} = 0. \tag{1.5}$$

Once these parameters, a and b, are evaluated, the precise nature of the stability can be ascertained. The results may be interpreted as whether $\lambda/\lambda_{cr} < 1$ or > 1 (i.e.) the postbuckling configuration carries more load or less than the critical load.

We see that in asymptotic postbuckling analysis, at a bifurcation point (point B in Fig. 1.3), the load–deflection behavior can be written from Eq. (1.3) as,

$$\lambda_m / \lambda_{cr} = (1 + a\varepsilon + b\varepsilon^2 + \ldots). \tag{1.6}$$

The possible outcome of the expression with only three terms considered in Eq. (1.6) is presented in Fig. 1.3. For the asymmetric case shown in Fig. 1.3, $b = 0$ in Eq. (1.6), and the expression becomes $\lambda/\lambda_{cr} = 1 + a\varepsilon$. We see that the post-buckling behavior is affected by the sign of the buckling displacement. Depending on the sign of 'a', the postbuckling behavior would be stable ($\lambda/\lambda_{cr} > 1$) or unstable ($\lambda/\lambda_{cr} < 1$). In other words, 'asymmetric' bifurcation emerges at the bifurcation point. In the second and third cases in Fig. 1.3, $a = 0$, and hence the bifurcation is symmetric, and the mode of postbuckling behavior varies with the sign of 'b'. According to Koiter (1945), if there is an initial imperfection in the structure of a very small magnitude $\overline{\varepsilon}$, then Eq. (1.6) may be rewritten by substituting $(\varepsilon - \overline{\varepsilon})$ for ε as

$$(1 - \lambda/\lambda_{cr})\varepsilon + a\varepsilon^2 + b\varepsilon^3 + \ldots = (\lambda/\lambda_{cr})\overline{\varepsilon} + \ldots. \tag{1.7}$$

Thus, we see that for a structure that is not geometrically perfect, one may obtain load–deflection curves from the parameters of the perfect structure. The imperfect paths are also shown in Fig. 1.3. In the asymmetric case, when $a\varepsilon < 0$, the maximum load capacity in the post-bifurcation state is reduced. The maximum load λ_m for $a < 0$ is shown to be (Dym 1974)

$$\lambda/\lambda_{cr} = 1 - 2(-a)^{1/2}\overline{\varepsilon}^{1/2}. \tag{1.8}$$

From Eq. (1.8), we see that the decrease in the maximum load is directly proportional to the imperfection parameter, $\overline{\varepsilon}$. A structure that exhibits such behavior is said to be 'imperfection sensitive', in which the imperfections cause a reduction in its load-carrying capacity. In the symmetric stable case where $a = 0$, $b > 0$, there is no reduction in the load due to initial imperfections. In the symmetric unstable case, where $a = 0$, $b < 0$, the structure is insensitive to imperfections, and the expression gives the maximum load.

$$\lambda/\lambda_{cr} = 1 - 2(-b/4)^{1/3}\overline{\varepsilon}^{2/3}. \tag{1.9}$$

Comparing Eqs. (1.8) and (1.9), the former is called a 'quadratic structure', and the latter is called a 'cubic structure' because of the difference in their exponents. From the above discussions, it is clear that a structure's 'total potential energy (TPE)' is sufficient to ascertain the type of instability the structure undergoes in the post-critical state. Table 1.1 shows structural systems and their stability behavior.

1.5 Tracing of Postbuckled Paths Using Energy Framework

The critical points lie on the primary load–deflection path of the structural system. This section discusses how these postbuckling paths can be traced. As stated earlier, the potential energy π is assumed to be linear in λ such that $\pi(\lambda, \mathbf{p}) = U(\mathbf{p}) - \lambda\, \mathbf{p}^{\mathrm{T}}\mathbf{q}$, where $U(\mathbf{p})$ represents the strain energy of the system. The vector $\mathbf{p}$ represents

Table 1.1 Some structures and their postbuckling behavior

Structure	Type of bifurcation
Column under axial load	Symmetric stable bifurcation
Circular ring under pressure	Symmetric stable bifurcation
Strut on unidirectional elastic foundation	Both symmetric stable and unstable bifurcation
Plate under axial load	Symmetric stable bifurcation
Cylindrical shell under axial load	Symmetric unstable bifurcation
Frames	Symmetric unstable bifurcation
Shallow arches	Snap-through buckling

the generalized displacements of the 'n' D.O.F. system, and λ $\mathbf{p}^{\mathrm{T}}\mathbf{q}$ represents the potential energy of the loading where $\mathbf{q}$ is the vector of total external loads. Once a value of λ has been prescribed, π represents a 'surface' in the $(n + 1)$ dimensional configuration space represented by the axes λ, $\mathbf{p}_i$ $(i = 1, 2, 3, \ldots n)$. One can imagine new energy surfaces for different but constant values of λ.

The equilibrium paths may be considered as loci of stationary points on the sets of energy surfaces. For example, for a simple single degree of freedom representation, the loci of energy surfaces for the three cases of stability paths are shown in Fig. 1.4 (Thomson and Hunt 1973). To represent them in an inverse form, the condition for static equilibrium of a structural system is that the energy function π $(\lambda, \mathbf{p}_i)$ is stationary with respect to all 'n' generalized coordinates simultaneously, which implies that $\pi_{,i}$ $(\lambda, \mathbf{p}_i) = 0$, where the comma represents the partial derivative of the preceding quantity with respect to the succeeding indices.

This can be written in an expanded form as

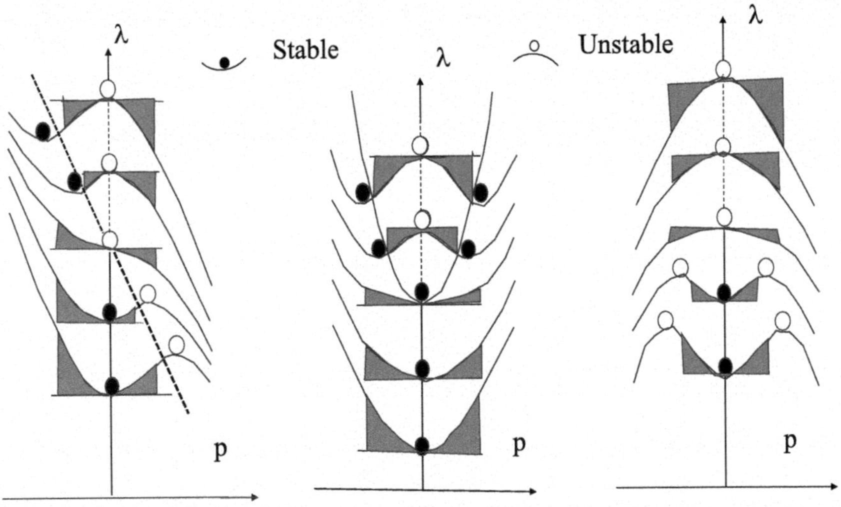

Fig. 1.4 Equilibrium paths as loci of energy surfaces

$$\pi_{,i} = 0 \quad i = 1, 2, 3 \ldots n. \tag{1.10}$$

Now expanding $\pi_{,i}$ using the Taylor series for an adjacent configuration $p_i + \delta p_i$, we get (in the indicial notation)

$$\pi_{,i}(p_i + \delta p_i) = \pi_{,i}(p_i) + \pi_{,ij}(p_i)\delta p_j + \frac{1}{2}\pi_{,ijk}(p_i)\delta p_j \delta p_k. \tag{1.11}$$

If $(p_i + \delta p_i)$ represents an equilibrium state, then it is necessary that $\pi_{,i}(p_i + \delta p_i) = 0$. Dropping the higher-order terms in the Taylor expansion of Eq. (1.11), we can write

$$\pi_{,i} + \pi_{,ij}\delta p_j = 0, \tag{1.12}$$

where the derivatives are evaluated at the point p_i. From the linearity of π, we can write $\pi_{,i} = U_{,i} - \lambda q_i$ and $\pi_{,ij} = U_{,ij}$. Substituting these expressions into the truncated Taylor expansion (Eq. 1.12) we get

$$U_{,ij}\delta p_j = \lambda q_i - U_{,i} \tag{1.13}$$

According to Castigliano's theorem, $U_{,i}$ is the force exerted on the structure in the p_i direction. Now the right-hand side term ($\lambda q_i - U_{,i}$) represents a vector 'r_i', which signifies the extent to which the equilibrium is satisfied or the unbalanced load vector. $U_{,ij}$ is the well-known 'tangent stiffness matrix', which relates the incremental displacements to incremental loads. The set of linear simultaneous equations arising out of Eq. (1.13) may be solved for $\delta\mathbf{p}$ and an improved estimate ($\mathbf{p} + \delta\mathbf{p}$) is obtained. At this stage, the unbalanced residual force vector '$\mathbf{r}$' is computed and applied repeatedly until '$\mathbf{r}$' becomes negligibly small. Using ($\mathbf{p} + \delta\mathbf{p}$) as the new starting point, the process may be repeated and solutions for various $\mathbf{q}$ levels are obtained. This procedure is called the 'Newton–Raphson' procedure, which works very well if the first approximation is sufficiently close to the solution and the equilibrium path is well-behaved (displacements monotonically increase with an increase in load). A brief discussion of several other path-following techniques is presented in the later part of this chapter.

It is seen that Eq. (1.13) is amenable to path following. Nevertheless, they do not provide any information on the nature of the critical points. During the solution of the nonlinear equations using $L.D.L^T$ factorization of the $U_{,ij}$ matrix, the pivotal coefficients c_i determine the stability of the system and are a convenient parameter to use numerically. One can use these stability coefficients to make a qualified statement about the nature of the stability of the equilibrium state. Thus, if a particular stability coefficient, c_i, is positive, it may be seen that the equilibrium state is stable with respect to the corresponding principal coordinate p_i. If the particular coefficient, c_j, is negative, then the equilibrium state is unstable with respect to the principal coordinate p_j. If c_k is zero, then we can say that the equilibrium state is critical with respect to the principal coordinate p_k. The absolute value of stability coefficients determines the degree of instability of the structure. Hence we see that once $U_{,ij}$ is formed at an equilibrium configuration, a simple examination of its properties reveals the state of all the corresponding stability coefficients and the structural system's stability.

1.6 Advanced Stability Investigations Using the Secant Matrix Technique

1.6.1 Total Potential Energy in Finite Element Analysis

The secant matrix method of geometrically nonlinear finite element analysis is based on the minimum of total potential energy (TPE) theorem. This method is based on the classical analysis of stability problems using Taylor expansion of the TPE at equilibrium (Koiter 1945; Thomson and Hunt 1973). It is seen from the literature (Thomson and Hunt 1973; Pignatro et al. 1995) that the TPE has always been the 'language' for solving structural stability problems. The numerical treatment of stability problems using total potential energy has been well established in the literature (Arboz and Hol 1990; Casciaro et al. 1992; Flores and Godoy 1992; Godoy et al. 1995) and books (El Naschie 1990; Waszczyszyn et al. 1994). The concept of the secant matrix technique is briefly presented in the following sections. Kroplin (1982) and his co-workers (Kroplin et al. 1985) proposed a secant matrix technique based on energy-based perturbation techniques for predicting the limit and bifurcation points for static and dynamic problems.

1.6.2 Description of Motion in Geometric Nonlinear Finite Element Analysis

In the kinematics of continua, there are two descriptions of motion called the (i) material or Lagrangian description (ii) spatial or Eulerian description. In the Lagrangian description, the motion of a particle in the continuum is expressed in terms of the coordinates in the reference of the earlier configuration. In the Eulerian description, attention is paid to a point in space and the particle that occupied the point is described in terms of its current coordinates. The Eulerian description is well suited for fluid mechanics problems. On the other hand, the Lagrangian description is commonly used in solid mechanics, and in the present study, the Lagrangian description is adopted in the geometric nonlinear finite element formulations.

In the Lagrangian description of motion, there are two different approaches, namely (a) the Total Lagrangian (T.L.) formulation and (b) the Updated Lagrangian (U.L.) formulation. The difference between these approaches lies in the choice of the reference configuration. In the T.L. description, all the static and kinematic variables are called the original undeformed configuration. In the U.L. description, the variables are referred to the last obtained configuration, i.e., the current configuration. In a general sense, both formulations are theoretically equivalent. The choice of the method has been debated very much in literature, but the answer to 'Which method is the best?' is still inconclusive. The answer is probably that the 'best method' choice varies from one problem to the other.

1.6.3 Secant Matrix Technique for the Analysis of Structural Stability

The basis of the secant matrix technique is explained briefly. Consider a deformable body at its undeformed position at time T_0. The deformed structure under the action of force **q** is shown in Fig. 1.5 at time $T_1 = T_0 + \Delta_t$. Using a Total Lagrangian (T.L.) description, the displacement of a material point P to P′ may be represented by a displacement vector $\mathbf{u} = \{u, v, w\}^{\mathrm{T}}$, where u, v, w are continuous, differentiable functions of initial coordinates (x, y, z), representing the component displacements in the Cartesian axes. Since the kinematic description is in T.L. format, Green's strain in engineering notation can be written as (using $\gamma_{xy} = 2\varepsilon_{xy}$)

$$\varepsilon = \{\varepsilon_x, \varepsilon_y, \varepsilon_z, \gamma_{xy}, \gamma_{yz}, \gamma_{zx}\}^{\mathrm{T}}. \tag{1.14}$$

These strain components may be related to the displacement gradients by the relation.

$$\begin{Bmatrix} \varepsilon_x \\ \varepsilon_y \\ \varepsilon_z \\ \gamma_{xy} \\ \gamma_{zy} \\ \gamma_{xz} \end{Bmatrix} = \begin{bmatrix} u_{,x} + \frac{1}{2}\left(u_{,x}^2 + v_{,x}^2 + w_{,x}^2\right) \\ v_{,y} + \frac{1}{2}\left(u_{,y}^2 + v_{,y}^2 + w_{,y}^2\right) \\ w_{,z} + \frac{1}{2}\left(u_{,z}^2 + v_{,z}^2 + w_{,z}^2\right) \\ \left(u_{,y} + v_{,x}\right) + \left(u_{,y}u_{,x} + v_{,y}v_{,x} + w_{,y}w_{,x}\right) \\ \left(v_{,z} + w_{,y}\right) + \left(u_{,z}u_{,y} + v_{,z}v_{,y} + w_{,z}w_{,y}\right) \\ \left(w_{,x} + u_{,z}\right) + \left(u_{,x}u_{,z} + v_{,x}v_{,z} + w_{,x}w_{,z}\right) \end{bmatrix}, \tag{1.15}$$

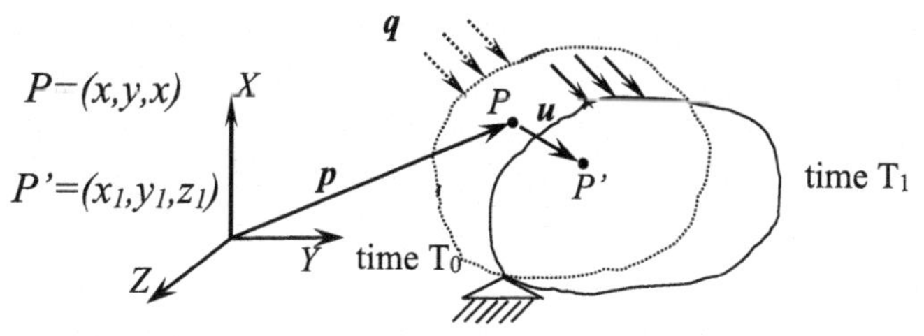

Fig. 1.5 Lagrangian description of the motion of a continuum

where the comma represents the differentiation of the preceding quantity with respect to the succeeding indices. For a linear elastic material, the physical relations between stress and strain may be written as $\boldsymbol{\sigma} = \mathbf{E}\,\boldsymbol{\varepsilon}$, where $\boldsymbol{\sigma}$ is the second Piola-Kirchhoff stress, which is the work conjugate to the Green's strain $\boldsymbol{\varepsilon}$. The increment in the specific strain energy at some point in the body for an infinitesimal change in displacements du, dv, dw could be written as

$$dU = \sigma_x d\varepsilon_x + \sigma_y d\varepsilon_y + \sigma_z d\varepsilon_z + \sigma_{xy} d\gamma_{xy} + \sigma_{yz} d\gamma_{yz} + \sigma_{zx} d\gamma_{zx}, \tag{1.16}$$

Thus, the work expended by the deformation of an initially rectangular elemental parallelopiped with edges dx, dy, dz referred to its original volume (in T.L. formulation, the integrals are carried out with respect to the undeformed volume) would be

$$U = \int_{\varepsilon=0}^{\varepsilon} (\sigma_x d\varepsilon_x + \sigma_y d\varepsilon_y + \ldots)\, dx\, dy\, dz. \tag{1.17}$$

Since the material of the body is assumed to be elastic, the integral in Eq. (1.17) is path-independent. The elastic constitutive matrix $\mathbf{E}$ relating stress and strain could be written as

$$\mathbf{E} = \frac{E}{(1+\nu)(1-2\nu)} \begin{bmatrix} (1-\nu) & \nu & \nu & 0 & 0 & 0 \\ & (1-\nu) & \nu & 0 & 0 & 0 \\ & & (1-\nu) & 0 & 0 & 0 \\ & \text{Symmetric} & & \frac{(1-2\nu)}{2} & 0 & 0 \\ & & & & \frac{(1-2\nu)}{2} & 0 \\ & & & & & \frac{(1-2\nu)}{2} \end{bmatrix} \tag{1.18}$$

where E is the Young's modulus and ν is the Poisson's ratio. Since the second Piola-Kirchhoff stress can be written as a product of the elasticity matrix and Green's strain, the strain energy is a function of the strain $\boldsymbol{\varepsilon}$ as $U(\boldsymbol{\varepsilon})$, which is indeed the strain energy of the elastic body per unit volume or in other words the specific strain energy. The function $U(\boldsymbol{\varepsilon})$ fully characterizes the elastic properties of the material of the body, using which one can get the stresses in the body for a given strain distribution and, conversely, the strains for a given set of stresses by the relation of Castigliano, $\boldsymbol{\sigma} = \partial U/\partial \boldsymbol{\varepsilon}$.

There are two forms of writing the strain energy U, namely U as a function of strains $U(\boldsymbol{\varepsilon})$ or in terms of the displacement gradients $U(\mathbf{g})$, where '$\mathbf{g}$' is the vector of displacement gradients appearing in Green's strain. For Eq. (1.15), '$\mathbf{g}$' could be written as

$$\mathbf{g} = \left\{u_{,x},\ u_{,y},\ u_{,z},\ v_{,x},\ v_{,y},\ v_{,z}, w_{,x},\ w_{,y},\ w_{,z}\right\}^{\mathrm{T}}. \tag{1.19}$$

The second form, the function of displacement gradients, is convenient for writing the equilibrium equations using the secant matrix technique. Given the first form, we can always write the second, but the converse is invalid. From Eq. (1.17), we see that the total value of strain components without initial strains can be written as a sum of ε^{L} that are linear in strain–displacement gradients and $\varepsilon^{\mathrm{NL}}$ are nonlinear function of the displacement gradients as

$$\boldsymbol{\varepsilon} = \boldsymbol{\varepsilon}^{\mathrm{L}} + \boldsymbol{\varepsilon}^{\mathrm{NL}}. \tag{1.20}$$

The total potential energy (TPE) can be written as

$$\pi = \frac{1}{2}\int_V \varepsilon^{\mathrm{T}}\mathbf{E}\varepsilon \, \mathrm{d}V - \int_\Gamma \mathbf{p}^{\mathrm{T}}\mathbf{q} \, \mathrm{d}\Gamma, \tag{1.21}$$

where $\mathbf{E}$ is the elasticity matrix, Γ is the portion on the structure where tractions are prescribed, and $\mathbf{q}$ and $\mathbf{p}$ are the force and displacement vectors, respectively. The strain energy component in Eq. (1.21) can be written using Eq. (1.20) as

$$U = \frac{1}{2}\int_V (\boldsymbol{\varepsilon}^{\mathrm{L}} + \boldsymbol{\varepsilon}^{\mathrm{NL}})^{\mathrm{T}}\mathbf{E}(\boldsymbol{\varepsilon}^{\mathrm{L}} + \boldsymbol{\varepsilon}^{\mathrm{NL}})\mathrm{d}V. \tag{1.22}$$

After carrying out the multiplication, this equation could be written symbolically into three parts as

$$U = \frac{1}{2}\int_V \mathbf{E}(\boldsymbol{\varepsilon}^{\mathrm{L}^T}\boldsymbol{\varepsilon}^{\mathrm{L}} + \left(\boldsymbol{\varepsilon}^{\mathrm{L}^T}\boldsymbol{\varepsilon}^{\mathrm{NL}} + \boldsymbol{\varepsilon}^{\mathrm{NL}^T}\boldsymbol{\varepsilon}^{\mathrm{L}}\right) + \boldsymbol{\varepsilon}^{\mathrm{NL}^T}\boldsymbol{\varepsilon}^{\mathrm{NL}}) \, \mathrm{d}V. \tag{1.23}$$

Equation (1.23) shows that the first expression in the above equation is quadratic, the second cubic, and the third quartic. Using the relation between the strains and displacement gradient in Eqs. (1.19) and (1.23), the above equation can be rewritten symbolically as a function of displacement gradients as

$$U(\mathbf{g}) = \int_V (U_2(\mathbf{g}) + U_3(\mathbf{g}) + U_4(\mathbf{g})) \, \mathrm{d}V. \tag{1.24}$$

The integrands in Eq. (1.24) have been grouped according to their order of dependence on the components of the displacement gradient vector as represented by the subscripts of the terms. In other words, $U_2(\mathbf{g})$, $U_3(\mathbf{g})$, $U_4(\mathbf{g})$ represent the monomials of $U(\mathbf{g})$ with the order of their degree equal to their subscripts.

For example, the $U_2(\mathbf{g})$ term could be written as

$$U_2(g) = \begin{Bmatrix} u_{,x} \\ v_{,y} \\ w_{,z} \\ u_{,y}+v_{,x} \\ v_{,z}+w_{,y} \\ w_{,x}+u_{,z} \end{Bmatrix}^T \overline{E} \begin{bmatrix} (1-\nu) & \nu & \nu & 0 & 0 & 0 \\ & (1-\nu) & \nu & 0 & 0 & 0 \\ & & (1-\nu) & 0 & 0 & 0 \\ & & & \frac{(1-2\nu)}{2} & 0 & 0 \\ & \text{Symmetric} & & & \frac{(1-2\nu)}{2} & 0 \\ & & & & & \frac{(1-2\nu)}{2} \end{bmatrix} \begin{Bmatrix} u_{,x} \\ v_{,y} \\ w_{,z} \\ u_{,y}+v_{,x} \\ v_{,z}+w_{,y} \\ w_{,x}+u_{,z} \end{Bmatrix}, \quad (1.25)$$

where $\overline{E} = \frac{E}{(1+\nu)(1-2\nu)}$.

After expansion, the $U_2(\mathbf{g})$ term could be written as

$$U_2(\mathbf{g}) = \overline{E}\begin{bmatrix} (1-\nu)u_{,x}^2 + \nu v_{,y}u_{,x} + \nu w_{,z}u_{,x} + \nu u_{,x}v_{,y} + (1-\nu)v_{,y}^2 + \nu w_{,z}v_{,y} \\ \nu u_{,x}w,z + \nu v_{,y}w_{,z} + (1-\nu)w_{,z}^2 + \frac{(1-2\nu)}{2}(u_{,y}+v_{,x})^2 \\ +\frac{(1-2\nu)}{2}(v_{,z}+w_{,y})^2 + \frac{(1-2\nu)}{2}(w_{,x}+u_{,z})^2 \end{bmatrix} \quad (1.26)$$

Inspection of Eq. (1.26) reveals that U_2 is a homogeneous expression in quadratic powers of the displacement gradients. Similarly, the cubic and quartic homogeneous expressions can also be written for $U_3(\mathbf{g})$ and $U_4(\mathbf{g})$.

For the finite element implementation, Eq. (1.24) can be rewritten as

$$U(\mathbf{g}) = \int_V (\mathbf{g}^{\mathrm{T}}\frac{1}{2}[\nabla^2 U_2]\mathbf{g} + \mathbf{g}^{\mathrm{T}}\frac{1}{6}[\nabla^2 U_3]\mathbf{g} + \mathbf{g}^{\mathrm{T}}\frac{1}{12}[\nabla^2 U_4]\mathbf{g})\, \mathrm{d}V \quad (1.27)$$

where ∇ is the gradient operator and $\nabla^2 U_n$ is the Hessian matrix of the monomials of U with degree $n = 2, 3, 4$ in Eq. (1.24). Equation (1.27) can be rearranged as:

$$U(\mathbf{g}) = \int_V \left(\mathbf{g}^{\mathrm{T}}\left(\frac{1}{2}[\nabla^2 U_2] + \frac{1}{6}[\nabla^2 U_3] + \frac{1}{12}[\nabla^2 U_4]\right)\mathbf{g}\right)\mathrm{d}V. \quad (1.28)$$

1.6.4 Finite Element Implementation of the Secant Matrix Technique

In the finite element method (F.E.M.) context, the integrals in Eq. (1.21) would be carried out for an assembly of finite elements. Using the finite element technique, the displacements u, v, w within a single element can be expressed as a sum of the product of nodal displacements vector **p** and the shape function $\boldsymbol{\varphi}$ as

$$\mathbf{u} = \varphi^{\mathrm{T}}\mathbf{p}, \quad (1.29)$$

where **p** is the vector of nodal degrees of freedom of a typical finite element. All the displacement derivatives can be expressed in terms of nodal degree of freedom vector **p**, for example, $\partial u/\partial x$ represented as $\mathrm{u}_{,x}$ as

$$\mathbf{u}_{,x} = \varphi_{,x}^{\mathrm{T}}\mathbf{p}. \quad (1.30)$$

Now, the displacement gradient vector is related to nodal displacements of a typical finite element by the relation:

$$\mathbf{g} = \mathbf{D}\mathbf{p}, \tag{1.31}$$

where $\mathbf{D}$ is a transformation matrix relating displacement gradients to nodal displacements. Now, applying the relation between the displacement gradients and the nodal degree of freedom of a typical finite element by using Eqs. (1.19), (1.29), and (1.31), we can recast Eq. (1.24) into a form amenable for finite element computation with the monomials of Eq. (1.24) as

$$U(\mathbf{p}) = \int_V \left(\overline{U}_2(\mathbf{p}) + \overline{U}_3(\mathbf{p}) + \overline{U}_4(\mathbf{p})\right) \mathrm{d}V, \tag{1.32}$$

where $\overline{U}_2(\mathbf{p})$, $\overline{U}_3(\mathbf{p})$, $\overline{U}_4(\mathbf{p})$ represent the monomials of $U(\mathbf{p})$ as a function of nodal displacements with the order of their degree equal to their subscripts.

As before, Eq. (1.32) can be rewritten as

$$U(\mathbf{p}) = \int_V \left(\mathbf{p}^{\mathrm{T}}\left(\left[\frac{1}{2}\left[\nabla^2\overline{U}_2\right] + \frac{1}{6}\left[\nabla^2\overline{U}_3\right] + \frac{1}{12}\left[\nabla^2\overline{U}_4\right]\right]\right)\mathbf{p}\right)\mathrm{d}V. \tag{1.33}$$

As this book follows, the integrals Eq. (1.33) can be carried out explicitly (Yang 1970) or by numerical integration. After carrying out the integration for a typical finite element, the total strain energy for a single element U^e can be written by combining Eqs. (1.28) and (1.29) as

$$U^e(\mathbf{p}) = \int_V \mathbf{p}^{\mathrm{T}}\left(\frac{1}{2}\mathbf{K}^{\mathrm{e}} + \frac{1}{6}\mathbf{N1}^{\mathrm{e}} + \frac{1}{12}\mathbf{N2}^{\mathrm{e}}\right)\mathbf{p}\,\mathrm{d}V. \tag{1.34}$$

Equation (1.34) coincides with Mallet-Marcal symbolism that $\left[\nabla^2\overline{U}_2\right]$ is a Hessian, which is independent of components of the displacement vector, and $\left[\nabla^2\overline{U}_3\right]$ and $\left[\nabla^2\overline{U}_4\right]$ are functions of first-order and second-order components of the displacement vector $\mathbf{p}$ and hence named **N1** and **N2**. The above form of writing the strain energy is called the Mallet and Marcal (1968) and Rajasekaran and Murray (1973) symbolism, the **N1**, **N2** formalism, or the secant matrix technique (Onate 1995) in the literature. The associated load vector is represented as $\mathbf{q}$. Now, Eq. (1.34), after carrying out the integration, the total potential energy can be written as

$$\pi^e(\mathbf{p}) = \mathbf{p}^{\mathrm{T}}\left(\frac{1}{2}\mathbf{K}^e + \frac{1}{6}\mathbf{N1}^e + \frac{1}{12}\mathbf{N2}^e\right)\mathbf{p} - \mathbf{p}^{\mathrm{T}}\mathbf{q}. \tag{1.35}$$

The equilibrium equations are obtained by applying the necessary condition of equilibrium $\partial\pi^{\mathrm{e}}/\partial\mathbf{p} = 0$ as

$$\left(\mathbf{K}^e + \frac{1}{2}\mathbf{N1}^e + \frac{1}{3}\mathbf{N2}^e\right)\mathbf{p} = \mathbf{q}. \tag{1.36}$$

Equation (1.36) is called the 'equilibrium equation' in the secant matrix formalism. After the assembly of the structural stiffness matrix, the matrices **K, N1, N2** are called the global 'secant matrices' and; hence Eq. (1.36) can simply be written as

$$\mathbf{K_S}\,\mathbf{p} = \mathbf{q}, \quad \text{where } \mathbf{K}_s = \mathbf{K} + \frac{1}{2}\mathbf{N1} + \frac{1}{3}\mathbf{N2}. \tag{1.37}$$

Since these matrices are used for finding the equilibrium of the structure and measuring the secant modulus in the load–displacement space, this technique can be taken to be known as the 'secant matrix technique'. Equation (1.37) can be solved by direct iteration procedure.

The equilibrium equation (Eq. 1.37) is valid in current and adjacent configurations. Hence, expanding Eq. (1.37) for $\mathbf{p} + \Delta\mathbf{p}$ and $\mathbf{q} + \Delta\mathbf{q}$ and taking only the first term in the Taylor expansion, we get

$$\Delta\mathbf{q} = \mathbf{K}\Delta\mathbf{p} + \mathbf{N1}\Delta\mathbf{p} + \mathbf{N2}\Delta\mathbf{p}, \tag{1.38}$$

or in other words

$$\Delta\mathbf{q} = \mathbf{K_T}\Delta\mathbf{p}, \text{ where } \mathbf{K}_T = \mathbf{K} + \mathbf{N1} + \mathbf{N2}. \tag{1.39}$$

Hence, in Eq. (1.39) we obtain the tangent stiffness matrices. Interestingly, the component matrices **K, N1, N2** repeat themselves in strain energy, secant, and tangent stiffness matrices. Another advantage of the secant matrix technique is that the same matrices can be used as both secant and tangent matrices by suitably varying the coefficients. This property of the secant matrices also gives a facility by which higher-order energy derivatives could be obtained, which is dealt with in further chapters. An essential application of the secant stiffness matrix is the direct computation of the residual force vector from the known value of displacement increments. This circumvents the stress computation for evaluating the residual force vector. The most important aspect of the secant matrix is that it combines the secant and tangent iteration algorithms in a unified manner. Nevertheless, one of the problems encountered in deriving secant stiffness matrices is that they may become non-unique and non-symmetrical unless special efforts are taken care to avoid them (Rajasekaran and Murray 1973).

1.6.5 *Possible Applications of the Secant Matrix*

There are three ways in which the secant matrices could be used to solve nonlinear system equations, as shown in Fig. 1.6. These are briefly explained below.

In the first case, the secant stiffness equation $\mathbf{K_s}\ (\mathbf{p}) = \mathbf{q}$ can be used as a 'total secant' equation to solve the displacements between the unloaded and the deformed configurations as

$$\mathbf{q} = [\mathbf{K_{Si}}]^{-1}\mathbf{p}, \tag{1.40}$$

where '*i*' is the iteration number. This procedure is a direct iterative procedure with limitations concerning the size of the load vector. The second approach is the 'incremental-total secant approach'. This is carried out by dividing the total applied force into several increments and obtaining displacement increments between each equilibrium configuration by using the total secant algorithm given by (Fig. 1.6b).

$$\mathbf{p_i^n} = [\mathbf{K_{Si}^{n-1}}]^{-1}\mathbf{q_i^n}, \tag{1.41}$$

where '*n*' is the load step number and '*i*' is the iteration number. The third approach is the incremental secant approach in which the tangent stiffness is used for traversing the equilibrium path and for ascertaining the unbalanced forces; the secant stiffness is used as shown in Fig. 1.6c, where $\mathbf{r_1}$, $\mathbf{r_2}$, $\mathbf{r_3}$ are the unbalanced residual forces. The solution procedure can be written as

$$\Delta\mathbf{p_i^n} = [\mathbf{K_{Ti}^{n-1}}]^{-1}\Delta\mathbf{q_i^n}, \tag{1.42}$$

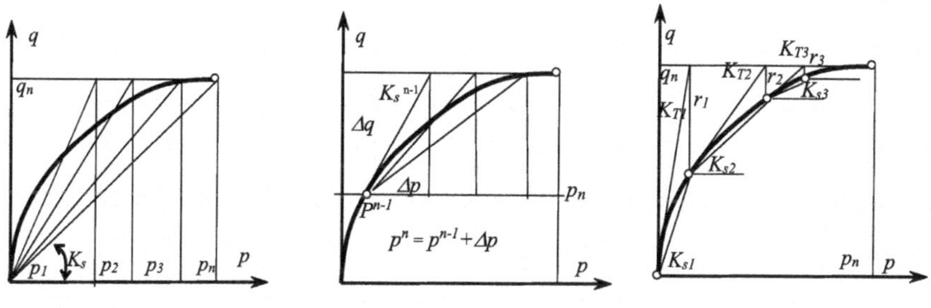

Fig. 1.6 Types of secant matrix approaches for geometrically nonlinear problems

1.7 Evolution of the Secant Matrix Technique

The secant matrix technique is widely used in nonlinear structural analysis, particularly postbuckling problems. Mallet and Marcal introduced this method in 1968, using a Total Lagrangian (TL) approach to derive secant stiffness matrices for displacement-based finite element analysis. Their formulation expressed the equilibrium equation as:

$$\left[\mathbf{K} + \frac{1}{2}\mathbf{N1} + \frac{1}{3}\mathbf{N2}\right]\mathbf{p} = \mathbf{q} \text{ or } [\mathbf{K}_s]\mathbf{p} = \mathbf{q}. \tag{1.43}$$

This approach, referred to as the 'N-notation' (Wood and Schrefler 1978), was an early attempt to handle nonlinearities. However, subsequent researchers, including Rajasekaran and Murray (1973), identified inadequacies in this method, particularly noting that the matrices **N1** and **N2** were not unique and lacked symmetry.

1.7.1 Historical Development and Improvements

Over the years, several authors have refined the secant matrix technique. Notably, Felippa (1974) introduced a parametric expression for the secant stiffness matrix, and later work by Felippa and colleagues (Felippa and Crivelli 1991; Felippa et al. 1994) reformulated the technique using a core congruential Total Lagrangian approach. This approach relies on the total potential energy as a function of displacement gradients.

1.7.2 Secant Matrix Technique in Nonlinear Structural Analysis

The study of buckling and structural stability is critical in addressing real-world engineering problems, as reflected in modern design codes like AS4100, EN 1993, and AISC. These codes integrate advanced analysis methods that account for stability, which is essential in contemporary structural design. With the growing reliance on computational tools such as finite element (F.E.) software, structural engineers can now analyze complex stability problems more efficiently. However, despite these advancements, a deep understanding of instability phenomena remains crucial to avoid misinterpreting the results of postbuckling analysis. Most F.E. packages still lack the necessary features to handle vital entities, such as accurately identifying critical points or equilibrium states.

The secant matrix technique has played a vital role in the postbuckling analysis of structures, primarily through the work of Gallagher (1969, 1974, 1984). Using perturbation methods, Gallagher applied the secant matrix approach to analyze structural instabilities. While most secant matrix formulations extend only to fourth-order energy terms, this technique's success depends heavily on the finite element method adopted.

Felippa's (1974) general parametric expression of the secant stiffness matrix was a significant contribution. The relationship between the secant matrix technique and the virtual work-based formulation (B-notation) was established by Wood and Schrefler (1978), with subsequent simplifications made by Badawi and Cuesens (1992). Onate (1995) and Onate et al. (2001) extended these formulations to predict instability points more accurately.

1.7.3 Secant Matrix Technique for Truss Structures

Secant matrix formulations have been widely applied in the analysis of truss structures. For large deflection analysis, Glabiz (1986) modified the secant matrix to include geometric stiffness. Narayanan and Krishnamoorthy (1994) developed secant matrices for nonlinear truss analysis using both updated and co-rotational formulations. Their work expanded the understanding of nonlinear behavior in truss structures. Onate et al. (2001) further refined these techniques, introducing incremental and total secant stiffness matrices for bar and solid elements, which were crucial for path tracing and instability analysis.

1.7.4 Secant Matrix Technique for Beams, Frames, and Thin-Walled Open Sections

The application of secant matrices in beam and frame structures has evolved significantly. Early work by Ebner and Ucciferro (1972) found Mallet and Marcal's (1968) secant matrix approach inadequate for large deflection problems in beams and frames. Later, Yang (1973) and Rajasekaran and Murray (1974) introduced more accurate formulations for postbuckling analysis. Morteza et al. (1997) used the secant matrix technique for the large deflection analysis of conventional plane beam structures and used the secant matrices of Rajasekaran and Murray (1973). For example, Ramesh and Krishnamoorthy (1993) reported that 'satisfactory results could not be obtained' when one element per member based on a quadratic secant matrix was used. They had to use six and twelve elements per member in their study to obtain good accuracy.

Szilard (1985) and Alexander and Churchill (1987) contributed to the nonlinear analysis of beams and frames by incorporating secant stiffness matrices. In the 1990s, researchers like Wong and Tin-Loi (1990) combined secant matrices with solution methods like Newton–Raphson and arc-length approaches for nonlinear frame analysis. Thin-walled structural elements pose unique challenges due to their sensitivity to torsional effects and instabilities. Rajasekaran and Murray (1974) developed secant matrices specifically for these types of structures, considering large displacements and torsion. Their work remains important in the nonlinear analysis of thin-walled beams with open cross-sections.

1.7.5 Secant Matrix Technique for Plates and Shells

Nevertheless, the conventional quadratic form of the secant matrices for beams presented in the literature (Mallet and Marcal 1968; Rajasekaran and Murray 1973; Alexander and Churchill 1987; Morteza et al. 1997) have been reported to be inadequate for large deformation problems. Palazotto et al. (1992) used the N1, N2 formulation to analyze the stability behavior of cylindrical laminated composite panels.

The secant matrix technique has also been applied to the postbuckling and large deflection analysis of plates and shells. Yang (1970, 1971) was among the first to use secant matrices for these problems. More research by Ganapathy and Varadhan (1995) and Balamurugan et al. (1996) extended these formulations to composite plates and shells, introducing geometric stiffness and mass matrices for dynamic instability analysis. Dawe and his co-workers (Dawe and Azizian 1987; Dawe et al. 1992, 1995; Lam and Dawe 1998) used the secant stiffness matrix in conjunction with the finite strip method for the postbuckling analysis of prismatic, isotropic, orthotropic, and laminated composite plates. Lam and Dawe (1998) used the secant stiffness matrix in conjunction with the finite strip method for the postbuckling analysis of prismatic, isotropic, orthotropic, and laminated composite plates. Kinematically exact modeling of beam deformations suggested by Reissner (1972) was used in the stability analysis of beams and frames using TL formulation (Gummadi and Palazotto 1998; Pai and Palazotto, 1996; Pajunen 1999).

1.8 Procedures for Solving Nonlinear Finite Element Equations

The solution procedures to trace the postbuckling paths beyond the critical points are called 'continuation methods or path tracing algorithms'. The solution procedure is like a 'black box' in many computer codes using the FE method. In the case of nonlinear FE computer programs, especially for postbuckling analysis, the user may encounter convergence issues unless the user is conversant with the solution procedure. The user must adopt the correct numerical technique for the problem. Generally, solution procedures for nonlinear FE equations may be classified in two ways. First, the solution methods are classified based on whether the unknowns are solved explicitly or implicitly, as shown in Table 1.2. Some of the implicit solution procedures are presented schematically in Fig. 1.7. The second way of classification, according to Bergan and Soreide (1973), in which solution procedures can be classified into three types, namely.

1. minimization techniques,
2. iterative techniques,
3. incremental techniques.

Table 1.2 Types of solution methods for nonlinear problems

Explicit methods	Implicit methods
Direct substitution method	Newton–Raphson(NR)
Conjugate gradient method	Line search method
Conjugate Newton method	Quasi-Newton method
Dynamic relaxation method	Newton Secant algorithm
	Arc-length method
	Minimum residual displacement method

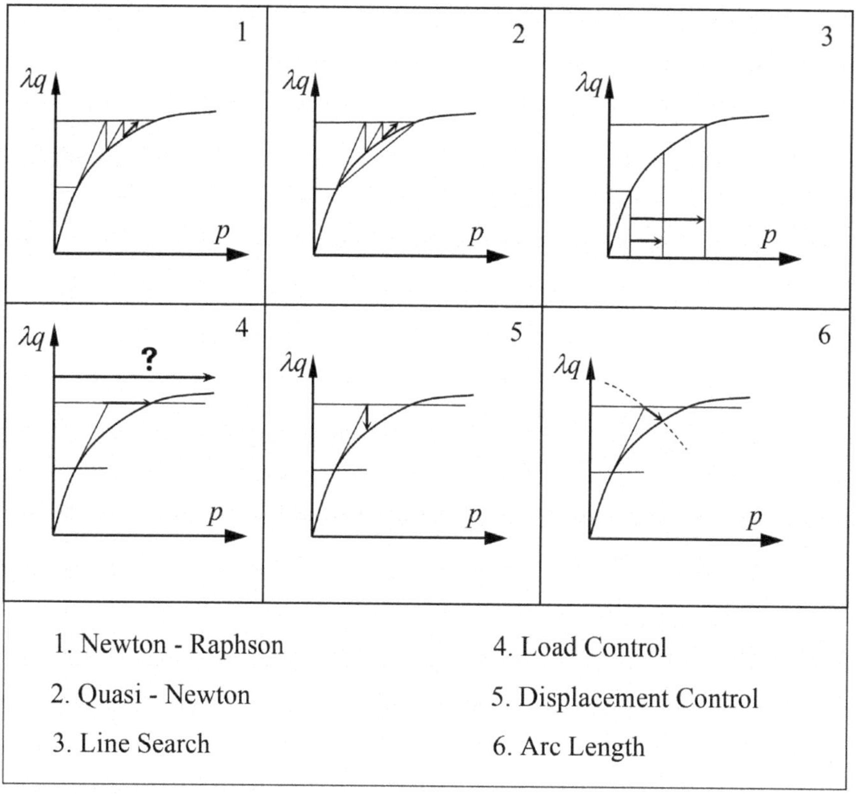

Fig. 1.7 Schematic representation of some implicit solution procedure

Bergan and Soreide (1973) presented an elaborate discussion of these methods. The solution procedures for the secant matrix technique fall in the implicit load control algorithm with or without Newton corrections. In the case of load control algorithms (Fig. 1.7(4)), we see that beyond the limit load, no solution is possible. Hence, we resort to displacement control algorithms (Fig. 1.7(5)). However, in snap-back problems, the displacement control algorithm fails, and recourse is made to the arc-length procedures (Fig. 1.7(6)).

1.8.1 Minimization Techniques

In structural stability theory, the elastic total potential energy is a function that attains a stationary value at equilibrium. For a discretized structure, these functionals may be written as

$$\pi = U(\mathbf{p}) - \mathbf{p}^{\mathrm{T}}\mathbf{q}. \tag{1.44}$$

The potential energy π consists of the total strain energy U of the body and the load potential, which is a product of displacement '**p**' and corresponding loads '**q**'.

The minimization techniques imply that the extreme values of the potential are sought by direct minimization techniques like the random search technique or Powell's conjugate direction method. The steepest descent method is the most straightforward of the gradient methods. These methods have been improved, and the Fletcher-Reeves method falls in this category. This method is graphically shown in Fig. 1.7(3).

1.8.2 Iterative Approaches

The following approach to the solution of nonlinear equations is obtained by setting the variation of the total potential energy to zero by simple equilibrium considerations. In the simplest form, these equations can be written as

$$\mathbf{K}_{\mathrm{S}}\mathbf{p} = \mathbf{q}. \tag{1.45}$$

$\mathbf{K}_{\mathrm{s}}$ is the secant stiffness matrix, and **p,q** are the displacements and loads. If $\mathbf{p}^*$ is a set of displacements that are not in equilibrium, then the residual unbalanced forces can be written as

$$\mathbf{r} = \mathbf{K}_{\mathrm{s}}\mathbf{p}^* - \mathbf{q}. \tag{1.46}$$

The equilibrium equations are usually solved using some iterative schemes, which, in a general form, can be written as

$$\mathbf{p}^{n+1} = \mathbf{p}^n - \lambda^n (\mathbf{A}^n)^{-1} (\mathbf{K}^n \mathbf{p}^n - \mathbf{q}), \tag{1.47}$$

where λ^n is the over-relaxation factor, and A^n is the nonsingular matrix used as a gradient for the nth iteration. For instance, the true gradient to the equilibrium relation corresponding to the potential's second derivative can generally be derived. By substituting the true gradient for $\lambda^n(A^n)^{-1}$ in Eq. (1.47), we get the well-known Newton–Raphson iterations. This method is quadratically convergent but expensive as it needs gradient computation for every iteration. On the other hand, if the same initial gradient is used during the entire iteration, we get the modified Newton–Raphson technique. Various norms, such as displacement, force, and energy norms, are used to terminate the iteration process.

1.8.3 Incremental Methods

The 'incremental methods' are techniques by which the nonlinear problem is solved in a stepwise incremental fashion. The incremental force–displacement relations can be written as a set of differential equilibrium equations

$$\mathbf{K}_{\mathrm{T}} \Delta \mathbf{p} = \Delta \mathbf{q}, \tag{1.48}$$

where $\mathbf{K}_{\mathrm{T}}$ is the tangent stiffness matrix. Incremental methods are divided into two groups. The first is the purely incremental method based on the differential equilibrium equation (Eq. 1.46). However, the solution may diverge from the true equilibrium path in these methods. The second group consists of methods that include measures that guide the solution back to the true path. The simplest pure incremental method is the Euler-Cauchy method. The most popular incremental–iterative technique uses a simple Euler-Cauchy increment followed by Newton–Raphson iteration.

The following sections explain the arc-length method, one of the robust path-following continuation methods.

1.8.4 Constraint Methods for Postbuckled Path Following

In traversing the limit point (L.P.), the conventional Newton–Raphson algorithms fail, as shown in Fig. 1.7(4). To overcome this, various constraint methods have been proposed in the literature. Some methods are normal plane or updated normal plane methods, the dominant displacement method, and the spherical and cylindrical arc-length methods. In all these constraint methods, the load level parameter λ is used to adjust the load level (i.e.) at any instant of loading, and the load vector can be written as

$$\mathbf{q}^n = \lambda^n \mathbf{q}, \tag{1.49}$$

where $\mathbf{q}$ is a fixed load vector and 'n' represents the load increment number. The constraint methods are the Newton–Raphson procedures with an added constraint to define the iterative load-deformation path. The constraint equations may be of spherical or cylindrical arc-length type (Crisfield 1981), the normal plane or updated normal plane methods (Riks 1979), the dominant displacement methods (DeBorst 1987; Rheinbolt 1981), and the residual energy suppression using cylindrical type constraints (Widijaja 1998). The general format of the iterative methods that adopt a flexible load level can be written as

$$\delta \mathbf{p}_i^n = \delta \overline{\mathbf{p}}_i^n + \delta \lambda_i^n \delta \hat{\mathbf{p}}_i^n, \tag{1.50}$$

where 'i' denotes the iteration number and 'n' denotes the load increment number. The vector $\delta \overline{\mathbf{p}}_i^n$ is the vector of iterative displacements related to the residual or unbalanced force vector that occurs in a normal Newton–Raphson method, which could be written as

$$\delta \overline{\mathbf{p}}_i^n = -\left[\mathbf{K}_{Ti}^n\right]^{-1} \mathbf{r}_i^n, \tag{1.51}$$

where $\mathbf{r}_i^n$ is the residual force vector computed at the current estimate of the total load parameter λ_{i-1}^n. The vector $\delta \hat{\mathbf{p}}_i^n$ is the iterative displacement vector corresponding to the total load vector '$\mathbf{q}$'. We can write the incremental displacement as

$$\Delta \mathbf{p}_i^n = \Delta \mathbf{p}_{i-1}^n + \delta \mathbf{p}_i^n, \tag{1.52}$$

and using Eq. (1.50), this can be written as

$$\Delta \mathbf{p}_i^n = \Delta \mathbf{p}_{i-1}^n + \delta \overline{\mathbf{p}}_i^n + \delta \lambda_i^n \delta \hat{\mathbf{p}}_i^n. \tag{1.53}$$

Inspection of Eq. (1.51) reveals that if the tangent stiffness matrix $\mathbf{K}_{\mathrm{Ti}}^n$ is evaluated, one can compute

The spherical constraint is introduced by specifying an arc length Δl, which is kept constant during an increment. This can be written as

$$\left[\Delta \mathbf{p}_i^n\right]^{\mathrm{T}} \Delta \mathbf{p}_i^n + \left[\Delta \lambda_i^n \mathbf{q}\right]^{\mathrm{T}} \Delta \lambda_i^n \mathbf{q} = \Delta \mathrm{l}^2 = \text{constant}. \tag{1.54}$$

The constraint equation can also be written in terms of a cylindrical surface as

$$\Delta \mathbf{p}_i^n = \Delta l^2 = \text{constant}. \tag{1.55}$$

By introducing Eq. (1.53) into Eq. (1.54) or Eq. (1.55), we get a quadratic equation in terms of $\delta\lambda$. Once the roots of the equation vis-à-vis $\delta\lambda$ is known, one may find the displacement vector

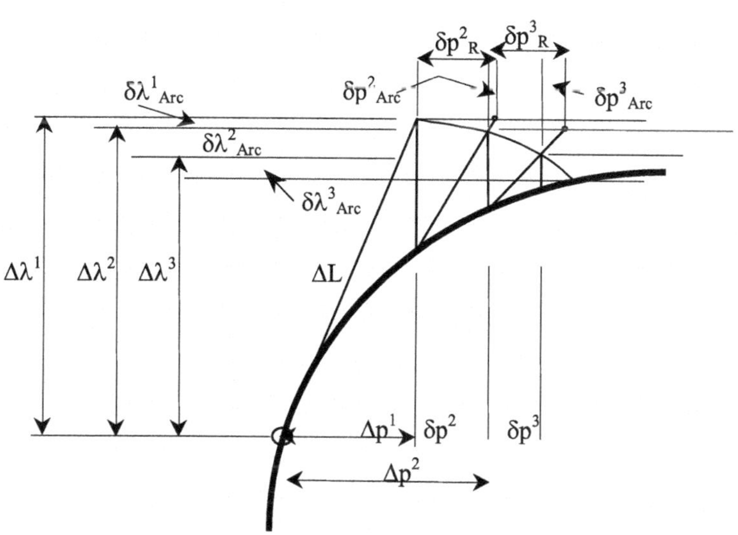

Fig. 1.8 Notations used in the **arc-length constraint method**

A root that avoids doubling back on the solution is to be selected out of the two roots. This is accomplished by selecting the root where the corresponding incremental displacement encircles the small angle (Crisfield 1981). This is written as

$$\angle\left(\Delta\mathbf{p}_i^n \Delta\mathbf{p}_{i-1}^n\right) = \text{min}. \tag{1.56}$$

Once $\delta\lambda_i^n$ is known then using Eq. (1.53), the $\Delta\mathbf{p}_i^n$ can be computed. There are also several variations to the arc-length methods (Fig. 1.8). For example, a unified presentation of Crisfield's and Rheinbolt's method was presented by Kouhia and Mikkola (1989). A survey on such solution methods is conducted by Crisfield and Shi (1991).

Instead of finding $\delta\lambda_i^n$ using the roots of the quadratic expression, Chan (1988) presented a simple and a novel expression by which $\delta\lambda_i^n$ can be computed by minimizing the unbalanced displacement norm as

$$\delta\lambda_i^n = (\delta\mathbf{p}_0 \cdot \delta\mathbf{p}/\delta\mathbf{p} \cdot \delta\mathbf{p}), \tag{1.57}$$

where $\delta\mathbf{p}_0$ and $\delta\mathbf{p}$ are the displacements due to the unbalanced force and the reference load, respectively. The minimum residual displacement (MRD) method suggested by Chan (1988) is very effective and circumvents the problems of computing the roots of the quadratic arc-length equation in which sometimes there may be a negative value inside the root.

1.9 Advanced Computational Methods for Path Tracing

The modern computational stability theory deals with path tracing, pinpointing, and path switching. Several schemes to trace the equilibrium path have been described, as mentioned earlier. Some stability indicators help ascertain important stability parameters during normal path tracing. For example, inspection of the number of negative pivots in the LDL^T decomposition of the $\mathbf{K}_{T.}$ and $\lambda\mathbf{q}$ may be used to check the closeness of the current equilibrium point to the stability point nearby. This is termed path measuring. Once a stability point is arrived, the singular point is identified as a limit or a bifurcation point. This is called pinpointing. After pinpointing, if a limit point is classified, the unstable equilibrium path may be traced using the path-tracing algorithm or continuation methods. If a bifurcation point is identified, then the solution path has to be switched, and this procedure is called 'path switching'. Mathematically, the condition determines the existence of a bifurcation in the nonlinear system.

$$\mathrm{Det}(\mathbf{K}_T) = 0 \text{ and } \mathbf{q}^T\phi = 0. \tag{1.58}$$

$\mathrm{Det}(\mathbf{K}_T)$ is the determinant of $\mathbf{K}_T$, and ϕ is the first eigenvector. To follow the bifurcated path, one can resort to the various branch-switching techniques as presented by Choong and Hangai (1993), Fujii et al. (1995), and Okazawa and Fujii (1996).

The most straightforward approach for branch switching is the eigenmode injection technique. The basic idea is to add a small perturbation to the present displacement field $\mathbf{p}_o$ at the bifurcation point

$$\mathbf{p}_n = \mathbf{p}_o + c\phi, \tag{1.59}$$

where 'c' is a constant $\boldsymbol{\phi}$ is the eigenvector. In a predictor–corrector format, the eigenmode injection method can be interpreted as replacing a normal tangent predictor ($\Delta\lambda$, $\mathbf{K}_T{}^{-1}\mathbf{q}$) by (0, $c\boldsymbol{\phi}$). It has been shown by Shi and Crisfield (1992) that this is a good predictor, both for symmetric and asymmetric bifurcations.

1.9.1 Stability Indicators

As the path tracing continues, an indicator is needed to reveal a critical point's approach. Shi and Crisfield (1992) suggested that $\mathrm{Det}(\mathbf{K}_{.T.})$, or the smallest pivot in the $L.D.L.^T$. factorization of the stiffness matrix, can be used as a stability indicator. The smallest pivot of $\mathbf{K}_{T.}$, available in many nonlinear FE codes, can be used as the stability indicator.

1.9.2 Automatic Load Incrementation

The choice of increment (load/displacement step) size is essential during path tracing. If it is too large, convergence does not occur or becomes slow. Hence, the increment size should reflect the degree of nonlinearity. Several methods have been presented in the literature. In particular, Bergan and Soreide (1978) proposed the current stiffness parameter (CSP) as a measure of the degree of nonlinearity. The current stiffness parameter $S_{P.}$ is given by

$$S_P = \frac{\left\{[\Delta \mathbf{q}]^i\right\}^T \left\{[\Delta \mathbf{p}]^1\right\}}{\left\{[\Delta \mathbf{q}]^1\right\}^T \left\{[\Delta \mathbf{p}]^i\right\}}, \tag{1.60}$$

where $\Delta \mathbf{p}^1$ and $\Delta \mathbf{p}^i$ are the increment in displacements in the 1st and the ith iterations and $\Delta \mathbf{q}^1$ and $\Delta \mathbf{q}^\mathrm{i}$ are the Euler load increments in the 1st and the ith iterations. The CSP is also a measure of the current tangent stiffness to the tangent stiffness at the beginning of the loading and is used to select the size of the next load increment. Later, Crisfield (1981) suggested another procedure to compute the next load increment as

$$\Delta\lambda_j = \Delta\lambda_{j-1}\frac{I_\mathrm{d}}{I_{j-1}}, \tag{1.61}$$

where I_d is the desired number of iterations, and I_{j-1} is the number of iterations required at the previous increment, which is of size $\Delta\lambda_{j-1}$. Ramm (1980) used the square root of $(I_\mathrm{d}/I_{\mathrm{j}-1})$, giving a smoother response. Bellini and Chulye (1987) used

$$\Delta\lambda_j = \sqrt[4]{\frac{I_2}{I_1}}\Delta\lambda_{j-1}, \tag{1.62}$$

where I_1 is the optimum number of iterations and I_2 is the number of iterations used in the previous increment.

1.10 Summary

This chapter introduces the elastic postbuckling behavior of structures, tracing the evolution of structural stability theory from classical works by Euler, Lagrange, and Koiter to modern computational approaches. The chapter discusses the significance of elastic stability in structures, particularly those prone to geometric nonlinearity, such as slender beams, plates, and shells. Key concepts such as critical points, bifurcation, and limit points are explored, highlighting their importance in understanding postbuckling behavior and utilizing reserve strength in design. It further elaborates on the mathematical formulation of stability using total potential energy (TPE) and the secant matrix technique for analyzing nonlinear behavior in structural systems. The chapter also reviews the formulation of the secant matrix technique for various structures. Finally, computational methods like path-following algorithms and the arc-length method are introduced for solving complex nonlinear equations in postbuckling analysis.

References

Alexander C, Churchill JE (1987) Nonlinear frame analysis by finite element methods. J Struct Eng 113(6):1221–1235

Arboz J, Hol JMAM (1990) Koiter's stability theory in a computer aided engineering environment. Int J Solids Struct 26:945–973. https://doi.org/10.1016/0020-7683(90)90011-J

Badawi M, Cusens AR (1992) Symmetry of the stiffness matrices for geometrically nonlinear analysis. Commun Appl Numer Methods 8:135–140

Balamurgan V, Ganapathy M, Varadhan TK (1996) Nonlinear dynamic instability of laminated composite plates using finite element method. Comput Struct 60:125–130. https://doi.org/10.1016/0045-7949(95)00368-1

Bellini PX, Chulye A (1987) An improved automatic incremental algorithm for the efficient solution of nonlinear finite element equations. Comp Struct 26:99–110. https://doi.org/10.1016/0045-7949(87)90240-9

Bergan PG, Soreide TH (1973) A comparative study of different numerical solution techniques as applied to nonlinear structural problems. Comput Methods Appl Mech Eng 2:185–201. https://doi.org/10.1016/0045-7825(73)90014-5

Bergan PG, Soreide TH (1978) Solution of large displacement and instability problems using current stiffness parameter. In finite elements in nonlinear mechanics, (eds.) P.G.Bergan et al, Tapir, Trondheim, 647–669

Casciaro R, Salerno G, Lanzo AD (1992) Finite element asymptotic analysis of slender elastic structures: A simple approach. Int J Numer Methods Eng 35:1397–1426. https://doi.org/10.1002/nme.1620350703

Chan SL (1988) Geometric and material nonlinear analysis of beam—columns and frames using the minimum residual displacement method. Int J Numer Methods Eng 26:2657–2669. https://doi.org/10.1002/nme.1620261206

Choong KK, Hangai Y (1993) Review on methods of bifurcation analysis for geometrically nonlinear structures. Bull Int Assoc Shell Spatial Struct 34:133–149. https://doi.org/10.1016/S0141-0296(96)00094-6

Crisfield MA (1981) A fast incremental/iterative solution procedure that handle snap-through. In: Computational methods in nonlinear structural and solid mechanics. 13:55–62. https://doi.org/10.1016/B978-0-08-027299-3.50009-1

Crisfield MA, Shi J (1991) A review of solution procedures and path following techniques in relation to the nonlinear finite element analysis of structures. In: Wriggers P, Wagner W (eds) springer Verlag, Berlin

Dawe DJ, Lam SSE, Azizian ZG (1992) Finite strip post local buckling analysis of composite prismatic plate structures. Comput Struct 48:1011–1123. https://doi.org/10.1016/0045-7949(93)90436-H

Dawe DJ, Wang S, Lam SSE (1995) Finite strip analysis of imperfect laminated composite plate under end shortening and normal pressure. Int J Num Meth In Engg 38:4193–4205

Dawe DJ, Azizian GZG (1987) Postbuckled stiffness of rectangular orthotropic composite laminates. In: Marshall H (ed) Composite structures, Elsevier, London. https://doi.org/10.1007/978-94-009-3455-9_10

Deborst R (1987) Computation of post bifurcation and post-failure behavior of strain softening solids. Comput Struct 25:211–224. https://doi.org/10.1016/0045-7949(87)90144-1

Dym CL (1974) Stability theory and its applications to structural mechanics. Noordhoff

Ebner AM, Uccifero JJ (1972) A theoretical and numerical comparison of elastic nonlinear finite element methods. Comput Struct 2:1043–1061. https://doi.org/10.1016/0045-7949(72)90055-7

Felippa C (1974) Discussion on the paper by Rajasekaran and Murray, (1973). J Struct Div 99:2521–2523. https://doi.org/10.1061/JSDEAG.0003667

Felippa C, Crivelli LA (1991) The core congruential formulation of geometrically nonlinear finite element. In nonlinear computational mechanics-state of the art, (eds.) P.Wriggers and W.Wagner, springer Verlag

Felippa C, Crivelli LA, Hugen B (1994) A survey of the core congruential formulation for geometrically nonlinear T.L. finite elements. Arch Comput Methods Eng 1:1–48. https://doi.org/10.1007/BF02736179

Flores FG, Godoy LA (1992) Elastic postbuckling analysis via finite element and perturbation techniques. Int J Numer Methods Eng 33:1775–1794. https://doi.org/10.1002/nme.1620330902

Flugge W (1972) Tensor analysis in continuum mechanics. Springer-Verlag, NJ. https://doi.org/10.1007/978-94-015-9988-7

Fujii F, Okazawa S, Gong XS (1995) Computational stability theory—its strategy. In: Proceedings of International Conference on Computer Engineering Science, Hawaii, USA. https://doi.org/10.1007/978-3-642-79654-8_247

Gallagher RH (1969) The finite element method in elastic stability analysis. Symposium on finite element techniques at the institut fur statik und dynamik der luft und raumfahrtkonstruktionnen, University of Stuttgart, Germany, 1–49

Gallagher RH (1974) Perturbation procedures in nonlinear finite element analysis. Proceedings of the Int. Conf. on Computational methods in nonlinear mechanics, Texas, 75–88

Gallagher RH (1984) Finite element elastic instability analysis. In: Symposium on Finite element methods in South Africa. CSIR, Pretoria

Ganapathy M, Varadhan TK (1995) Nonlinear free flexural vibrations of laminated circular cylindrical shells. Compos Struct 30:33–49. https://doi.org/10.1016/0263-8223(94)00025-5

Glabiz (1986) The influence of second-order values in the numeric analysis of bar structures apt to snap through. In: Yagawa G, Atluri SN (ed) Proceedings of International Conference on Comput Methods Computational Mechanics ‘86

Godoy LA, Barbero EJ, Raftoyiannis IG (1995) Finite elements for postbuckling analysis I—the ‘W’ formulation. Comput Struct 56:1009–1017. https://doi.org/10.1016/0045-7949(94)00572-K

Gummadi LNB, Palazotto AN (1998) Large strain analysis of beams and arches undergoing large rotations. Int J Nonlinear Mech 33:615–645. https://doi.org/10.1016/S0020-7462(97)00033-4

Koiter WT (1945) Elastic stability and postbuckling behaviour. Doctoral Dissertation at Technische Hooge School, Delft, Netherlands also as Technical report TRFFDL-TR-70-25, Wright Patterson Airforce Base, 1970

Kouhia M, Mikkola M (1989) Tracing the equilibrium path beyond simple critical points. Int J Numer Methods Eng 28:2932–2941. https://doi.org/10.1002/nme.1620281214

Kroplin B (1982) Postbuckling instability analysis of shells using mixed method. In Buckling of shells, Ed.E.Ramm, Univ Stuttgart 175–192

Kroplin B, Dinkler D, Hillman J (1985) An energy perturbation method applied to nonlinear structural analysis. Comput Methods Appl Mech Eng 52:885–897. https://doi.org/10.1016/0045-7825(85)90019-2

Lam SSE, Dawe DJ (1998) Nonlinear analysis of rectangular laminated plates under end shortening using finite strip method. Leondes CT, Gordon

Mallet RH, Marcal PV (1968) Finite element analysis of nonlinear structures. J Struct Div 94:2081–2105. https://doi.org/10.1061/JSDEAG.0002066

Malvern LE (1969) Introduction to the mechanics of a continuous medium. Prentice-Hall, Hoboken, NJ

Mejia-Nava RA, Imamovic I, Hajdo E, Ibrahimbegovic A (2022) Nonlinear instability problem for geometrically exact beam under conservative and non-conservative loads. Eng Struct 265:114446. https://doi.org/10.1016/J.Enggstruct.2022.114446

Morteza AM, Torkamani, Sonmez M, Cao J (1997) Second order elastic plane frame analysis using finite element method. J Struct 123:1225–1235. https://doi.org/10.1061/(ASCE)0733-9445(1997)123:9(1225)

Narayanan G, Krishnamoorthy CS (1994) Investigations on truss element models for the static response of geometrically nonlinear structural systems. J Struct Eng 21:145–163

Naschie EI (1990) Stress, stability and chaos in structural engineering: an energy approach. McGraw-Hill, London, New York

Okazawa S, Fujii F (1996) Global local procedures to compute the stability point. In: Third Asian-Pacific Conference on Computational Mechanics, Seoul, Korea

Onate E (1995) On the derivation and possibilities of the secant stiffness matrix for nonlinear finite element analysis. Comput Mech 15:572–593. https://doi.org/10.1007/BF00350269

Onate E, Tschope H, Wriggers P (2001) Combination of the critical displacement method with a damage model for structural instability analysis. Eng Comput 18:642–662. https://doi.org/10.1108/02644400110387190

Pai PF, Palazotto AN (1996) Large deformation analysis of flexible beams. Int J Solids Struct 33:1335–1353. https://doi.org/10.1016/0020-7683(95)00090-9

Pajunen S (1999) Fully nonlinear analysis of plane frame using kinematically exact beam elements. In: Mikkola M, Kouhia R (eds) NSCM-12 Proceedings Helsinki

Palazotto AN, Chien LS, Taylor WW (1992) Stability characteristics of laminated cylindrical panels under transverse loading. AIAA J 30:1649–1653. https://doi.org/10.2514/3.11113

Pignatro M, Rizzi, Luango A (1995) Stability of bifurcation and post critical behaviour of elastic structures. Elsevier Publishers, New York, NY. https://doi.org/10.1016/C2009-0-12669-7

Rajasekaran S, Murray DW (1973) Incremental finite element matrices. J Struct Eng 99:2423–2438. https://doi.org/10.1061/JSDEAG.0003667

Rajasekaran S, Murray DW (1974) Finite element large deflection analysis of thin walled beams of open section. Proceedings of International Conference on Finite Element Methods in Engineering, (eds). Pulmono V.A and Kabaila A.P, 435–454

Ramesh G, Krishnamoorthy CS (1993) Postbuckling analysis of structures by dynamic relaxation. Int J Numer Methods Eng 36:1339–1364. https://doi.org/10.1002/nme.1620360806

Ramm E (1980) Strategies for tracing the nonlinear response near limit points. In: Wunderlich W (ed) Nonlinear finite element analysis in structural mechanics. Springer, Berlin. https://doi.org/10.1007/978-3-642-81589-8_5

Reissner E (1972) On one-dimensional finite-strain beam theory, the plane problem. J Appl Math Phys 23:795–804. https://doi.org/10.1007/BF01602645

Rhienbolt WC (1981) Numerical analysis of continuation methods for a nonlinear structural problem. Comput Struct 13:103–113

Riks E (1979) An incremental approach to the solution of snapping and buckling problems. Int J Solids Struct 15:529–551

Santos HAFA (2012) A complementary-energy based criterion for the stability analysis of geometrically exact framed structures. Comput Struct 106–107:196–203. https://doi.org/10.1016/j.compstruc.2012.05.004

Shi J, Crisfield MA (1992) A simple indicator and branch switching technique for hidden unstable equilibrium paths. Finite Elem Anal Des 12:303–312. https://doi.org/10.1016/0168-874X(92)90039-F

Szilard Rudolph (1985) Critical load and postbuckling analysis by FEM using energy balancing technique. Comput Struct 20(1–3): 277–286

Thompson JMT, Hunt GW (1973) A general theory of elastic stability. John Wiley and Sons, Hoboken, NJ. https://doi.org/10.1016/0022-5096(63)90003-6

Waszczyszyn Z, Chichon C, Radwanska M (1994) Stability of structures by finite element methods. Elsevier, Amsterdam

Widijaja R, Buddi (1998) Path following technique based on residual energy suppression for nonlinear finite element analysis. Comput Struct 66(2–3), 201–209

Wong MB, Tin-Loi F (1990) Geometrically nonlinear analysis of elastic framed structures. Comput Struct 3:66–640. https://doi.org/10.1016/0045-7949(90)90241-S

Wood RD, Schrefler B (1978) Geometrically nonlinear analysis—a correlation of finite element notations. Int J Numer Methods Eng 12:635–642. https://doi.org/10.1002/nme.1620120408

Yang TY (1973) Matrix displacement solution to the elastica problems of beams and frames. Int J Solids Struct 9:829–842. https://doi.org/10.1016/0020-7683(73)90006-1

Yang TY (1970) Flexible plate finite element on elastic foundation. J Struct Engg ASCE 96:2083–2101

Yang TY (1971) A finite element procedure for large deflection analysis of plates with initial deflections. AIAA J 9:468–1473

Chapter 2
A Co-rotation-Based Secant Matrix Procedure for the Elastic Postbuckling Analysis of Truss Structures

Abstract This section introduces a co-rotational (CR)-based secant matrix procedure for conducting elastic postbuckling analysis of truss structures. The co-rotational approach reduces the complexity of nonlinear problems by isolating rigid body rotations and translations and considers only the local deformations of the truss members. This chapter explains the implementation of the secant matrix method onto a CR framework. Several numerical examples of different truss systems are presented. In this chapter, individual member postbuckling expressions are derived using which interaction problems of member and global buckling are solved. Although some problems use synthetic data, they provide much-needed information on the underlying postbuckling phenomenon.

Keywords Co-rotational framework · Secant matrix · Elastic postbuckling · Truss instability

2.1 Reticulated Shell Structures

Structural systems composed of discrete steel elements, such as geodesic domes and reticulated shells, are popular due to their effectiveness and economy covering large spaces. These systems are widely used in architectural and industrial applications because they efficiently combine structural integrity with lightweight design. Continued research has been conducted on the computational methods for analyzing the static nonlinear response of reticulated structures. A crucial aspect to consider in large reticulated structures is that instability may occur from member buckling or global buckling independently or in a combined way.

In the following sections, we derive geometric nonlinear finite element formulations for space truss elements using a secant matrix technique within a co-rotational framework.

A. Jayachandran, *Elastic Postbuckling Analysis of Structural Systems*, Solid Mechanics and Its Applications 232, https://doi.org/10.1007/978-3-031-89936-2_2

2.2 Co-rotational Procedure in Geometric Nonlinear Analysis of Truss Structures

Two widely used approaches in nonlinear finite element analysis are the Total Lagrangian (TL) and Updated Lagrangian (UL) formulations. Both effectively analyze large deformation problems where significant structural displacements and rotations occur. However, an alternative approach, the co-rotational (CR) formulation, has been developed to enhance computational accuracy and efficiency by isolating rigid body motion from local deformation. This distinction is achieved by introducing a local coordinate system that moves and rotates with the element. In this system, rigid body movements are represented by nodal displacement components set to zero, allowing the remaining natural or deformational displacements to be efficiently isolated.

Once the rigid body displacements are excluded from the CR framework, the remaining strain-displacement relationships can be treated using either the Total Lagrangian (TL) or Updated Lagrangian (UL) procedures. This leads to the development of hybrid CR-TL and CR-UL formulations, which leverage the strengths of the co-rotational framework and member strain-displacement relations.

The number of natural degrees of freedom (DOF) in the CR formulation is crucial, representing the minimal set of geometric variables required to describe the element's deformation. The CR formulation proves efficient in space trusses, which typically exhibit five rigid body modes (three translations and two rotations) and one deformational mode. Hence, the CR approach is particularly well-suited for nonlinear analysis of space truss structures.

2.3 Some Previous Work on Postbuckling of Reticulated Structures

Research on the nonlinear and postbuckling behavior of truss structures has evolved considerably. Early foundational work by Meek (1971), Hangai and Kawamata (1972), and Noor (1974) laid the groundwork for understanding nonlinear truss behavior. Further advancements in computational methods came from Cichon (1984), Papadrakakis (1987), and Zhu et al. (1993), who contributed significantly to the analysis of large reticulated structures under nonlinear conditions.

Yang and Leu (1990) and Crisfield (1991) provided critical reviews on the geometric nonlinear behavior of trusses, while Narayanan and Krishnamoorthy (1994) explored postbuckling responses in compression members. These studies highlighted the significance of Euler-type instabilities in truss systems, which had previously been modeled by Usama Rustom Madi (1984).

Researchers such as Kondoh and Atluri (1985) and Tanaka et al. (1985) developed numerical methods for addressing member instabilities in both plane and space trusses. Murtha Smith (1994) introduced refinements to improve the accuracy of post-buckling analysis. Additional contributions by Punniakotty et al. (1994), Yang and Yang (1996), and Blanford (1995) expanded the understanding of static and dynamic postbuckling behavior, taking into account member yielding and instability.

The co-rotational (CR) technique, first proposed by Wempner (1969) and later extended by Belytschko and Hsieh (1973), became a key method in truss analysis. It provided an efficient way to separate rigid body motions from local deformations. Argyris et al. (1976) introduced the concept of 'natural displacements', further enhancing the CR method. Matiasson and Samuelsson (1984) then systematized the development of CR-TL and CR-UL formulations, establishing a robust framework for large deflection and postbuckling analysis. The CR technique continues to be a reliable method for analyzing the nonlinear behavior of space trusses.

2.4 Description of Motion in the Co-rotational Formulation

The motion of a space truss element within the co-rotational formulation is depicted in Fig. 2.1. The figure illustrates the element in its reference configuration at time $T = 0$, its co-rotated configuration at time $T = t$, and its neighboring configuration at time $T = t + \Delta t$. For clarity, the displacements experienced by node 1 from the reference to the co-rotated configuration are shown.

If x, y, z represent the coordinates of a material point in the reference configuration, then x', y', z' represent the coordinates of the same point in the co-rotated configuration. The unit vector $\mathbf{e}_x$ describes the motion of the truss element in the co-rotated system. Within this co-rotating frame, the deformation of the element to

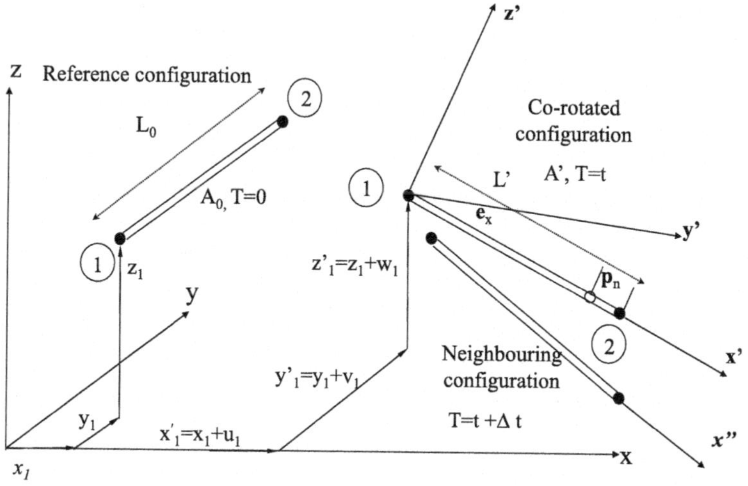

Fig. 2.1 Space truss element in its reference and co-rotated coordinates

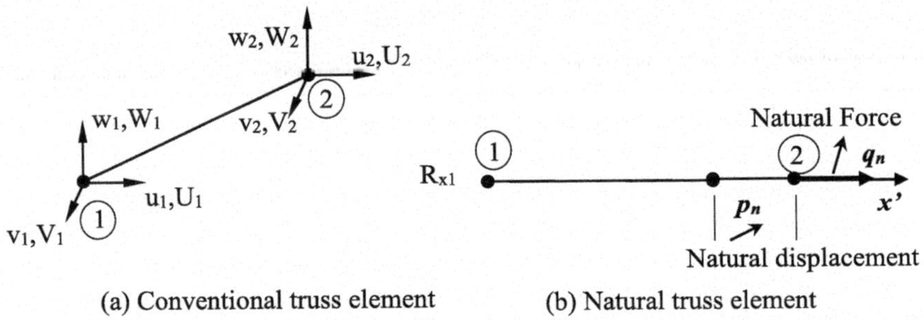

Fig. 2.2 Conventional and natural truss elements

an adjacent or neighboring configuration can be described using either a TL or UL format, depending on the specific requirements of the analysis.

Figure 2.2a shows a traditional 3D truss element with global displacements and the related nodal forces. Figure 2.2b illustrates the space truss element within a co-rotational framework. This co-rotated representation is created by setting the local coordinate system's origin at node 1 in the deformed state, as outlined in Fig. 2.1. The location of node 2 is recalculated based on the truss element's base vector in both its undeformed (initial) and deformed (current) configurations.

For a space truss element, only axial strain—extension or contraction along the element's axis—induces deformation. The remaining degrees of freedom (three translational and two rotational) result from rigid body motion, affecting the element's position and orientation but not its state of strain.

Two main steps are needed to construct a geometric nonlinear finite element model for a space truss element in the co-rotational framework: (a) mapping the reference coordinates to the co-rotated system and (b) linking the co-rotated system to the element's strain-inducing deformations.

The first step derives nonlinear transformations to convert global variables into co-rotated variables. The second step establishes the natural force-displacement relationships through the natural tangent stiffness matrix, allowing for geometric nonlinear effects and the element's deformations.

As per the polar decomposition theorem in continuum mechanics, the processes of separating rigid body motion from material deformation are fully independent. This principle makes isolating the pure deformation from the rigid body rotation possible, particularly for large deformations and rotations. With this decomposition, the next step is constructing both the tangent and secant stiffness matrices. These matrices capture the structure's response to loading as they establish the connection between forces and displacements in geometric nonlinearity. The tangent stiffness matrix accounts for incremental changes and is used in iterative solutions. The secant stiffness matrix provides the basis for establishing equilibrium at any point in loading. Together, these matrices are embedded into the co-rotational approach, enabling accurate analysis of nonlinear behavior in truss structures.

2.5 Nonlinear Kinematics in the Co-rotated State

In the co-rotational framework, the origin of the co-rotating coordinate system (x', y', z') is positioned at node 1 of the deformed element, with the x' axis extending through node 2, as shown in Fig. 2.1. The y' and z' axes are perpendicular to x'. However, their precise definitions are unnecessary for the co-rotational (CR) truss element. Within the finite element context, let the global displacement and force degrees of freedom (DOFs) at nodes 1 and 2 be represented as $\mathbf{p} = \{u_1,\ v_1,\ w_1,\ u_2,\ v_2,\ w_2\}^{\mathrm{T}}$ and $\mathbf{q} = \{U_1,\ V_1,\ W_1,\ U_2,\ V_2,\ W_2\}^{\mathrm{T}}$, respectively. The natural force and displacement are represented by single DOF quantities, $\mathbf{q}_{\mathrm{n}}$ and $\mathbf{p}_{\mathrm{n}}$, as shown in Fig. 2.2b.

The truss element's initial and co-rotated configuration lengths are denoted by L_0 and L', respectively. Likewise, the cross-sectional areas are indicated as A_0 and A', respectively, as shown in Fig. 2.1. The co-rotating base vector $\mathbf{e}'_x$ referenced to the global coordinate system can be expressed as $\mathbf{e}'_x = \frac{1}{L'}\left\{x'_{21},\ y'_{21},\ z'_{21}\right\}$, where $x'_{21} = x'_2 - x'_1$ and so on. The current length L' is computed as $L' = \left(\left(x'_{21}\right)^2 + \left(y'_{21}\right)^2 + \left(z'_{21}\right)^2\right)^{1/2}$.

$$\mathbf{p}' = \mathbf{A}\mathbf{p}. \tag{2.1}$$

For an increase in $\mathbf{p}$ by $\delta\mathbf{p}$ in Eq. (2.1), the following relations may be used.

$$\delta\mathbf{p}' = \mathbf{A}\delta\mathbf{p}. \tag{2.2}$$

Assuming that the truss element deforms into a nearby configuration, indicated by overbar quantities at $t + \Delta t$, the increment in natural displacement, $\delta\mathbf{p}_{\mathrm{n}}$, is expressed by $\overline{x}_2 - x'_2$. The updated length, $\overline{x}_2$, can be calculated as $\overline{x}_2 = \overline{\mathrm{e}}_x^{\mathrm{T}}\overline{X}_{21}$, where $\overline{\mathrm{e}}_x^{\mathrm{T}}$ is the co-rotating base vector in the adjacent configuration and $\overline{X}_{21} = \left\{\overline{x}_{21},\ \overline{y}_{21},\ \overline{z}_{21}\right\}$. This can be rewritten as $\delta\mathbf{p}_{\mathrm{n}} = \overline{L} - L$. The increments in natural displacement and those in the current coordinates are thus related as described.

$$\left[\delta\mathbf{p}_{\mathrm{n}}\right] = [-1,\ 0,\ 0,\ 1,\ 0,\ 0]\begin{Bmatrix} \delta u'_1 \\ \delta v'_1 \\ \delta w'_1 \\ \delta u'_2 \\ \delta v'_2 \\ \delta w'_2 \end{Bmatrix} \tag{2.3}$$

This can be rewritten as $\delta\mathbf{p}_{\mathrm{n}} = \mathbf{H}\,\delta\,\mathbf{p}'$. In the general co-rotational context, $\mathbf{H}$ is a matrix, but in this formulation, this is represented by a boldface type even though it is a vector. Therefore, the natural and global variables can be related by combining Eqs. (2.2) and (2.3) as $\delta\mathbf{p}_{\mathrm{n}} = \mathbf{H}\ \mathbf{A}\ \delta\ \mathbf{p}$.

2.6 Transformation of Nodal Forces

The global force vector, $\mathbf{q}$, and the co-rotated force vector, $\mathbf{q}'$, are connected through the transformation matrix $\mathbf{A}$ as follows:

$$\mathbf{q} = \mathbf{A}^{\mathrm{T}}\mathbf{q}'. \tag{2.4}$$

Since node 1 is located at the origin of the co-rotated system and aligned with the x' axis of this system, the reaction force at node 1, $\mathbf{R}_{x1}$ is determined by equilibrium as $\mathbf{R}_{x1} = -\mathbf{q}_{\mathrm{n}}$. Thus, the relationship between the co-rotated (or local) forces and the natural force, $\mathbf{q}_{\mathrm{n}}$ is established accordingly.

$$\begin{Bmatrix} U_1' \\ V_1' \\ W_1' \\ U_2' \\ V_2' \\ W_2' \end{Bmatrix} = \begin{Bmatrix} -1 \\ 0 \\ 0 \\ 1 \\ 0 \\ 0 \end{Bmatrix} q_{\mathrm{n}}. \tag{2.5}$$

Equation (2.5) can be rewritten as:

$$\mathbf{q}' = \mathbf{H}^{\mathrm{T}}\mathbf{q}_{\mathrm{n}}. \tag{2.6}$$

This equation forms the core of the co-rotational approach. By applying the incremental operator to Eq. (2.6), we obtain:

$$\delta\mathbf{q}' = \mathbf{H}^{\mathrm{T}}\delta\mathbf{q}_{\mathrm{n}} + \delta\mathbf{H}^{\mathrm{T}}\mathbf{q}_{\mathrm{n}}. \tag{2.7}$$

The first term on the right-hand side represents the change in forces within the co-rotated system resulting from the variation in the natural force $\mathbf{q}_{\mathrm{n}}$. This term can be derived from the natural stiffness relation $\delta\mathbf{q}_{\mathrm{n}} = \mathbf{K}'\delta\mathbf{p}_{\mathrm{n}}$, where $\mathbf{K}'$ is the natural stiffness matrix, a scalar value since there is only one degree of freedom in the space truss element. The second term on the right-hand side of Eq. (2.7) accounts for the variation in forces due to the rotation of the natural force vector while its magnitude remains unchanged. This result aligns with the polar decomposition theorem from continuum mechanics, which indicates that changes in a deformable body's force system can be represented as a combination of 'stretch' (first term) and 'rotation' (second term) transformations or vice versa.

A vital aspect of the co-rotational approach is evaluating the vector $\mathbf{H}^{\mathrm{T}}$. In this study, $\delta\mathbf{H}^{\mathrm{T}}$ is derived using geometric principles. The vector $\delta\mathbf{H}^{\mathrm{T}}$ comprises two components, expressed as $\delta\mathbf{H}^{\mathrm{T}} = \delta(\mathbf{H}_{\mathrm{t}})^{\mathrm{T}} + \delta(\mathbf{H}_{\mathrm{r}})^{\mathrm{T}}$, where $\mathbf{H}_{\mathrm{t}}$ and $\mathbf{H}_{\mathrm{r}}$ represent the translational and rotational contributions, respectively.

As depicted in Fig. 2.3, the truss element in the co-rotated configuration undergoes displacements $\{\delta u_1', \delta v_1', \delta w_1'\}$ at node 1 and $\{\delta u_2', \delta v_2', \delta w_2'\}$ at node 2 (not shown

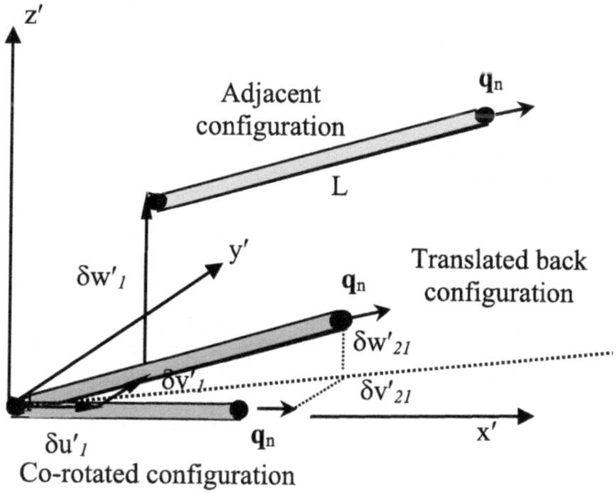

Fig. 2.3 Change in the co-rotated force vector from geometric principles

in Fig. 2.3 for clarity). To assess the change in co-rotated forces, we translate the adjacent configuration back by subtracting the displacements $\{\delta u_1,\ \delta v_1, \delta w_1\}$ from both nodes 1 and 2. This is referred to as the 'translated back configuration'.

At node 2, this translation corresponds to a displacement of $\{\delta u'_2 - \delta u'_1, \delta v'_2 - \delta v'_1, \delta w'_2 - \delta w'_1\}$ which, following our previous convention, can be expressed as $\{\delta u'_{21},\ \delta v'_{21}, \delta w'_{21}\}$. This translation does not induce any change in the co-rotated forces. Therefore, $\delta(\mathbf{H}_t)^T$ is a null vector, and it follows that any variation in the vector **H** in Eq. (2.3) or Eq. (2.5), which are translation-dependent, also yields a null vector.

We employ a reverse rotation to determine the change in the co-rotated forces, moving the natural force vector from the 'translated back' configuration to the co-rotated configuration. This reverse process entails rotating the force about the z' axis, followed by a rotation about the y' axis. Based on geometric considerations, this rotation produces a force change of $\left(\mathbf{q}_n \delta w'_{21}/L\right)$ along the z' axis and $\left(\mathbf{q}_n \delta v'_{21}/L\right)$ along the y' axis.

Thus, $\delta(\mathbf{H}_r)^T$, which represents the variation in the co-rotated force due to the rotation of the natural force as well as reaction-induced changes, can be formulated as

$$\delta\left(\mathbf{H}_r^T\right) = \begin{Bmatrix} \delta U'_1 \\ \delta V'_1 \\ \delta W'_1 \\ \delta U'_2 \\ \delta V'_2 \\ \delta W'_2 \end{Bmatrix} = \frac{\mathbf{q}_n}{L} \begin{Bmatrix} 0 \\ -\delta v'_{21} \\ -\delta w'_{21} \\ 0 \\ \delta v'_{21} \\ \delta w'_{21} \end{Bmatrix}. \tag{2.8}$$

Since $\delta \mathbf{H}^T = \{0\} + \delta(\mathbf{H}_r)^T$, the right-hand side of Eq. (2.8) can be expressed as $\delta \mathbf{H}^T \mathbf{q}_n$. It is seen that Eq. (2.8) aligns with the formulation presented by Mattiasson and Samuelsson (1984) based on continuum mechanics principles. All necessary terms in Eq. (2.7) have been derived at this stage.

By applying a variation to Eq. (2.4) and using Eq. (2.7), we obtain:

$$\delta \mathbf{q} = \mathbf{A}^{\mathrm{T}}\mathbf{H}^{\mathrm{T}}\delta \mathbf{q}_{\mathrm{n}} + \mathbf{A}^{\mathrm{T}}\delta \mathbf{H}^{\mathrm{T}}\mathbf{q}_{\mathrm{n}}. \tag{2.9}$$

The vector $\delta \mathbf{H}^{\mathrm{T}}$ can be expressed in co-rotated quantities by introducing a matrix $\mathbf{B}$ such that $\delta \mathbf{H}^{\mathrm{T}} = \mathbf{B}\delta \mathbf{p}'$ where $\mathbf{B}$ is given by

$$\mathbf{B} = \begin{bmatrix} 0 & 0 & 0 & 0 & 0 & 0 \\ 0 & 1 & 0 & 0 & -1 & 0 \\ 0 & 0 & 1 & 0 & 0 & -1 \\ 0 & 0 & 0 & 0 & 0 & 0 \\ 0 & -1 & 0 & 0 & 1 & 0 \\ 0 & 0 & -1 & 0 & 0 & 1 \end{bmatrix}. \tag{2.10}$$

Using the relation $\delta \mathbf{q}_{\mathrm{n}} = \mathbf{K}'\delta \mathbf{p}_{\mathrm{n}}$ and Eq. (2.3), Eq. (2.9) can be rewritten as

$$\delta \mathbf{q} = \mathbf{A}^{\mathrm{T}}\mathbf{H}^{\mathrm{T}}\mathbf{K}'\mathbf{H}\delta \mathbf{p}' + \mathbf{A}^{\mathrm{T}}\mathbf{q}_{\mathrm{n}}\mathbf{B}\delta \mathbf{p}'. \tag{2.11}$$

Using Eq. (2.2), Eq. (2.11) can be written in terms of global coordinates as

$$\delta \mathbf{q} = \mathbf{A}^{\mathrm{T}}\mathbf{H}^{\mathrm{T}}\mathbf{K}'\mathbf{H}\mathbf{A}\delta \mathbf{p} + \mathbf{A}^{\mathrm{T}}\mathbf{B}\mathbf{A}\delta \mathbf{p}'\mathbf{q}_{\mathrm{n}}. \tag{2.12}$$

Now, Eq. (2.12) can be recast into tangent stiffness relations in the global coordinate as

$$\delta \mathbf{q} = \mathbf{A}^{\mathrm{T}}\left(\mathbf{H}^{\mathrm{T}}\mathbf{K}'\mathbf{H} + \mathbf{q}_{\mathrm{n}}\mathbf{B}\right)\mathbf{A}\delta \mathbf{p}. \tag{2.13}$$

The global tangent stiffness matrix can be written after carrying out the multiplication in Eq. (2.13) as

$$\mathbf{K}_{\mathrm{T}} = \mathbf{A}^{\mathrm{T}}\left(\mathbf{H}^{\mathrm{T}}\mathbf{K}'\mathbf{H} + \mathbf{q}_{\mathrm{n}}\mathbf{B}\right)\mathbf{A} = \begin{bmatrix} \mathbf{K}_{\mathrm{C}} & -\mathbf{K}_{\mathrm{C}} \\ -\mathbf{K}_{\mathrm{C}} & \mathbf{K}_{\mathrm{C}} \end{bmatrix}. \tag{2.14}$$

The different terms of $\mathbf{K}_{\mathrm{C}}$ are obtained as

$$\mathbf{K}_{\mathrm{C}} = \left(\mathbf{K}' - \mathbf{q}_{\mathrm{n}}/L\right)\begin{bmatrix} A_{11}^2 & A_{11}A_{12} & A_{11}A_{13} \\ A_{12}A_{11} & A_{12}^2 & A_{12}A_{13} \\ A_{13}A_{11} & A_{13}A_{12} & A_{13}^2 \end{bmatrix} + \mathbf{q}_{\mathrm{n}}/L\begin{bmatrix} 1 & 0 & 0 \\ 0 & 1 & 0 \\ 0 & 0 & 1 \end{bmatrix}. \tag{2.15}$$

The values of transformation matrices A_{11}, A_{12}, and A_{13}, which relate to the local and global coordinates, can be found in McGuire et al. (2015).

2.7 Introduction of Secant Matrix Formalism into Co-rotational Framework

In Eq. (2.13), $\mathbf{K}'$ denotes the natural or local tangent stiffness matrix derived from local or co-rotated coordinates. This study employs a Total Lagrangian (TL) approach, utilizing the secant matrix method to formulate the local tangent stiffness matrix. The local tangent stiffness matrix is a scalar since the system features a single degree of freedom. Following the notation of Mallet and Marcal (1968), Green's strain is split into linear and nonlinear parts, represented as $\varepsilon_i = \varepsilon_i^L + \varepsilon_i^{\mathrm{NL}}$. For a truss element in its natural coordinates, this can be expressed as $\varepsilon = p_{\mathrm{n},x} + \frac{1}{2} p_{n,x}^2$, where $p_{\mathrm{n},x} = \frac{\partial p_{\mathrm{n}}}{\partial x}$.

Consequently, the strain energy due to natural deformation is given by:

$$U = \frac{1}{2} \int_{\Omega} \left(\mathbf{p}_{\mathrm{n},x} + \frac{1}{2} \mathbf{p}_{\mathrm{n},x}^2 \right) C \left(\mathbf{p}_{\mathrm{n},x} + \frac{1}{2} \mathbf{p}_{\mathrm{n},x}^2 \right) \mathrm{d}\Omega. \tag{2.16}$$

For the truss element, where $C = EA_0$ (with E as Young's modulus and A_0 as the cross-sectional area in the original state), we get:

$$U = \frac{C}{2} \int_0^L \left(\mathbf{p}_{\mathrm{n},x}^2 + \mathbf{p}_{\mathrm{n},x}^3 + \frac{1}{4} \mathbf{p}_{\mathrm{n},x}^4 \right) \mathrm{d}x'. \tag{2.17}$$

The term $\mathbf{p}_{\mathrm{n},x}$ can be represented in the finite element context as $p_{\mathrm{n},x} = \phi_{,x} \mathrm{p}_{\mathrm{n}}$, where the shape function ϕ is given by $\phi = x'/L_0$. Applying the shape function and integrating, we find $\mathbf{p}_{\mathrm{n},x} = \mathbf{p}_{\mathrm{n}}/L_0$. Using this expression and including the loss of potential due to external loads, the total potential energy π becomes:

$$\pi = \frac{C}{2} \int_0^{L_0} \left(\mathbf{p}_{\mathrm{n},x}^2 + \mathbf{p}_{\mathrm{n},x}^3 + \frac{1}{4} \mathbf{p}_{\mathrm{n},x}^4 \right) - \mathbf{p}_{\mathrm{n}} \mathbf{q}_{\mathrm{n}}. \tag{2.18}$$

According to the principle of minimum potential energy, $\partial \pi / \partial \mathbf{p}_{\mathrm{n}} = 0$. Following the approach of Rajasekaran and Murray (1973), the equilibrium equation becomes:

$$\mathbf{q}_{\mathrm{n}} = \frac{\mathbf{E}\mathbf{A}_0}{\mathbf{L}_0} \left(\mathbf{1} + \frac{3\mathbf{p}_{\mathrm{n}}}{\mathbf{L}_0} + \frac{3\mathbf{p}_{\mathrm{n}}^2}{2\mathbf{L}_0^2} \right) \mathbf{p}_{\mathrm{n}}. \tag{2.19}$$

This equation represents the secant stiffness matrix relation, which, in the notation of Mallet and Marcal (1968), can be written as:

$$\mathbf{q}_{\mathrm{n}} = \left(\mathbf{K} + \frac{\mathbf{N1}}{\mathbf{2}} + \frac{\mathbf{N2}}{\mathbf{3}} \right) \mathbf{p}_{\mathrm{n}}, \tag{2.20}$$

where **K** is the linear stiffness matrix, **N1** is the first-order incremental matrix, and **N2** is the second-order incremental matrix. The global nodal force vector can then be expressed as:

$$\mathbf{q} = \mathbf{H} \cdot \mathbf{A} \cdot \mathbf{q}_{\mathrm{n}}. \tag{2.21}$$

In the Mallet and Marcal (1968) notation, the tangent stiffness matrix is derived by expanding the equilibrium Eq. (2.20) using a Taylor series up to the first order, yielding:

$$\Delta \mathbf{q}_{\mathrm{n}} = (\mathbf{K} + \mathbf{N1} + \mathbf{N2})\Delta\ \mathrm{p}_{\mathrm{n}}, \tag{2.22}$$

where Δ denotes the Euler incremental operator. Therefore, in the present context, the 'natural' tangent stiffness relation, derived using the secant matrix approach, is:

$$\delta \mathbf{q}_{\mathrm{n}} = \frac{EA_0}{L_0}\left(1 + \frac{6\mathrm{p}_{\mathrm{n}}}{L_0} + \frac{9\mathrm{p}_{\mathrm{n}}^2}{2L_0^2}\right)\delta \mathbf{p}_{\mathrm{n}}. \tag{2.23}$$

Thus, in the co-rotational secant matrix method, the $\mathbf{K}'$ matrix is given by:

$$\mathbf{K}' = \frac{EA_0}{L_0}\left(1 + \frac{6\mathbf{p}_{\mathrm{n}}}{L_0} + \frac{9\mathbf{p}_{\mathrm{n}}^2}{2L_0^2}\right). \tag{2.24}$$

2.8 Influence of Individual Member Instability

Individual members are prone to local and global buckling in a structural assembly, such as space trusses or reticulated shells. Many practical struts in space trusses have symmetric cross-sections about two perpendicular axes and exhibit high torsional stiffness, often buckling about the weaker axis. This local bifurcation buckling of struts substantially affects the structure's subsequent load transfer and overall postbuckling behavior. Thus, incorporating the effect of member buckling is essential in the elastic postbuckling analysis of truss structures.

The postbuckling stiffness of a pin-ended axially loaded strut has been described by Britvec (1973, 1995) as $\mathrm{d}Q_{\mathrm{A}}/\mathrm{d}\Delta = (Q_{\mathrm{E}}/2L_0)$, where Q_{E} is the Euler critical load, and Q_{A} and Δ represent the applied load and axial displacement, respectively. This relation was utilized by Kondoh and Atluri (1985), Tanaka et al. (1985), and Kondoh and Tanaka (1986) to model compression member instability in a UL formulation. Yang and Yang (1996) included member instability effects through the plastification of cross-sections. In this study, the individual struts in space frames are assumed to be pin-ended and axially loaded. Under these assumptions, the postbuckling stiffness of each strut is determined as follows.

The initial imperfection of a strut with length L is given by:

$$w_0 = \delta_0 \sin\ (\pi x/L), \tag{2.25}$$

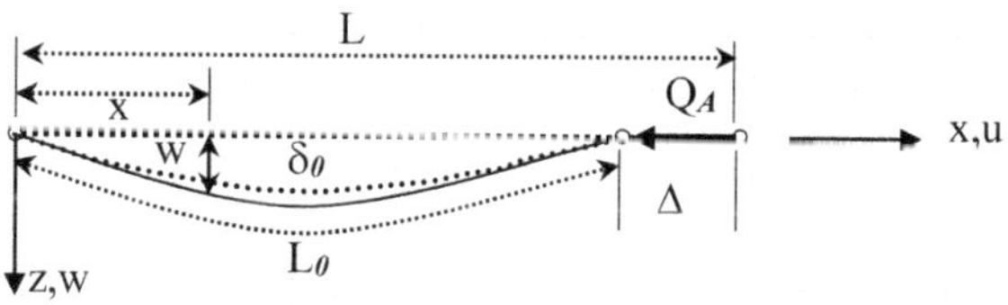

Fig. 2.4 Deformed shape of an initially imperfect strut

where δ_0 is the amplitude of the initial imperfection. The total deflection of the strut can be expressed as:

$$w = \delta \ \sin\ (\pi x/L) = \left[\delta_0/(1 - Q_A/Q_E)\right] \sin(\pi x/L), \tag{2.26}$$

where Q_A is the applied axial load, and Q_E is the Euler critical load. The length of the curved strut in its deformed state is shown in (Fig. 2.4):

$$L_0 = \int_0^L \sqrt{(1 + (\mathrm{d}w/\mathrm{d}x)^2}\, \mathrm{d}x. \tag{2.27}$$

The axial shortening due to the bowing effect can be written as $\Delta_b = L - L_0$. Since Eq. (2.27) cannot be solved explicitly by assuming a quadratic relation between the curved length and chord length, the integral in Eq. (2.27) can be approximated as

$$L = \frac{L_0}{1 + 2/3(2\delta_0/L_0)^2}. \tag{2.28}$$

The total end shortening Δ due to both the axial and bowing effect can be written as $\Delta = \Delta_a + \Delta_b$ which is simply

$$\Delta = \frac{Q_A\, L_0}{EA} + L_0 - \frac{L_0}{1 + 2/3(2\delta_0/L_0)^2}. \tag{2.29}$$

From this relation of Q and Δ the postbuckling stiffness of an imperfect column can be obtained as

$$\frac{\mathrm{d}Q_A}{\mathrm{d}\Delta} = \frac{1}{\left[\frac{L_0}{EA} + \frac{L_0}{Q_A}\left(1 - \frac{1}{1+2/3(2\delta/l)^2}\right)\right]} \tag{2.30}$$

in which δ is given by

$$\delta = \frac{\delta_0}{1 - {Q_A}/{Q_E}}. \tag{2.31}$$

The initial imperfection in the present case has been assumed as $0.001L_0$. When Eq. (2.30) is used as the postbuckled stiffness for the strut, the automatic load incrementation procedure, solving with arc-length methods, becomes more straightforward. Nevertheless, when Q_A approaches the value of Q_E, the strut's postbuckling stiffness becomes negligible; hence, small increments near the Euler critical load are unavoidable. Hence, the present study implemented a procedure of scaling down the loading increments near the Euler critical load of individual members.

2.9 Nonlinear Postbuckling Analysis of Truss Structures

The formulation discussed earlier has been integrated into a specialized computational program incorporating multiple solution methods, including the conventional Newton-Raphson procedure, the arc-length method, and the minimum residual displacement (MRD) technique detailed in Sections 1.8 and 1.9 of Chap. 1. Automatic load incrementation is also employed, based on the 'current stiffness parameter (CSP)' concept introduced by Bergan and Soreide (1973) and further improved by Bergan and Soreide (1978), ensuring stability in the solution process during nonlinear analyses.

To validate and calibrate both the formulations, five truss structure problems were selected. The first of these, illustrated in Fig. 2.5, is a two-bar toggle structure characterized by highly nonlinear behavior. Results from the co-rotational Total Lagrangian (CR-TL) method were compared with those from a standard TL approach, with a summary presented in Table 2.1. The comparison shows that the CR formulation significantly reduces the number of iterations required for convergence, demonstrating its efficiency and effectiveness for geometrically nonlinear truss structures.

2.9.1 Asymmetric Bifurcation in Truss Systems

The second problem dealt with a truss structure exhibiting asymmetric bifurcation behavior, similar to that studied by Crisfield (1997). The example uses synthetic data, and the details are shown in Fig. 2.6. The example involves applying a small

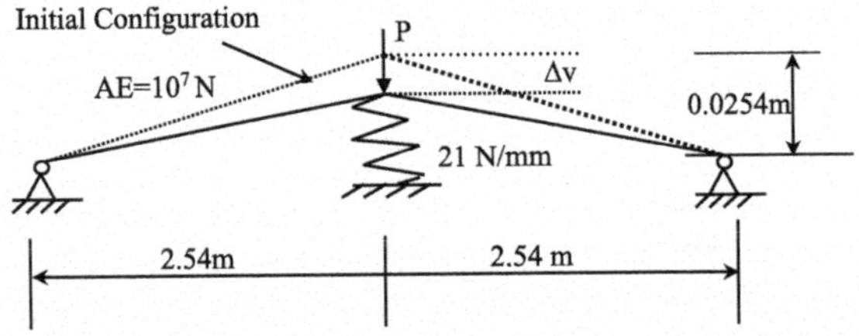

Fig. 2.5 Two-bar truss-I

Table 2.1 Comparison of CR-TL and TL procedures for two-bar truss-I

	Load P Newton's	TL procedure Narayanan and Krishnamoorthy (1994)		Equations (2.22) and (2.24) (CR-TL)	
		Number of iteration	Apex displacements Δv(mm)	Number of iterations	Apex displacements Δv(mm)
1	27.2	3	0.23	3	0.23
2	54.5	5	1.00	4	0.99
3	81.8	5	1.76	5	1.76
4	109.0	7	2.00	3	2.00
5	136.3	9	2.16	2	2.16
6	163.6	9	2.28	2	2.28
7	190.9	8	2.39	2	2.39
8	218.1	7	2.49	2	2.49
9	254.5	6	2.59	2	2.59
10	272.7	5	2.65	2	2.65
Total number of iterations		64	–	27	–

vertical load disturbance at node '2', which produces two distinct bifurcation paths. The resulting bifurcation behavior is depicted in Fig. 2.7, where the horizontal displacement at node '3' is plotted against the load factor.

The analysis further explores vertical displacements at node '2' (Fig. 2.8), revealing that both bifurcation paths share the same slope at the bifurcation point, verifying the accuracy of the present formulation. The CSP, discussed in Section 1.9.2 of Chap. 1, was plotted against the load factor in Fig. 2.9. A notable kink occurs at the bifurcation point, where CSP reaches its maximum, indicating potential oscillations due to the sharp snap-back behavior seen in Fig. 2.7.

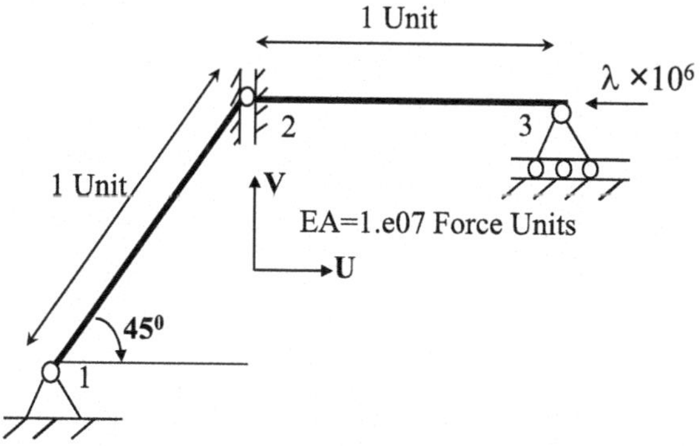

Fig. 2.6 Two-bar truss-II exhibiting asymmetric bifurcation

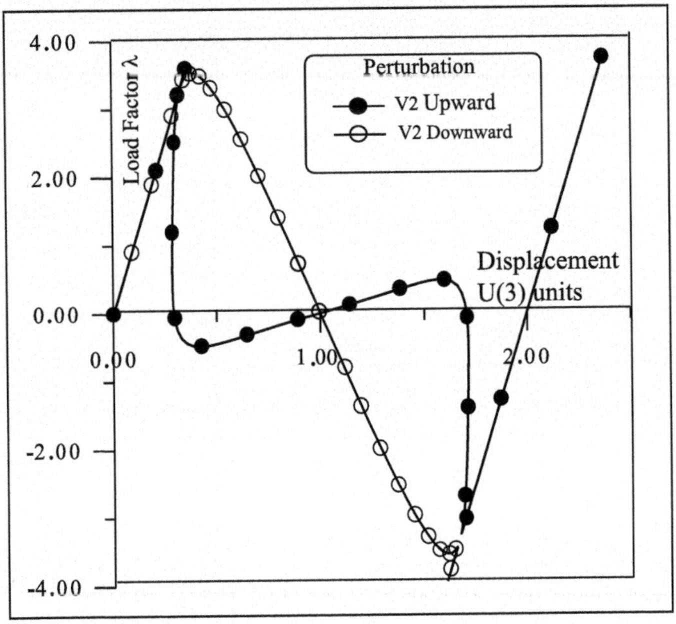

Fig. 2.7 Load vs. deformation U_3—two-bar truss-II

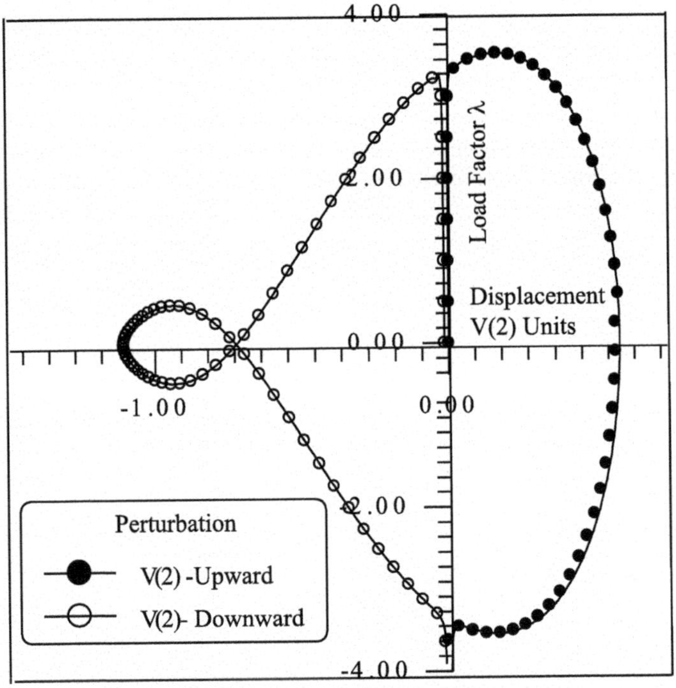

Fig. 2.8 Load versus deformation V2—two-bar truss-II

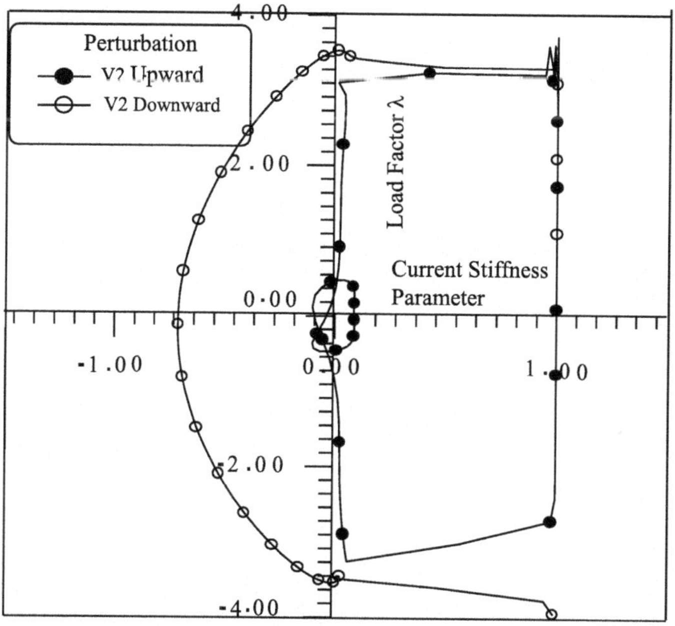

Fig. 2.9 Load versus current stiffness parameter—two-bar truss-II

The conventional arc-length method, however, struggled to handle these oscillations, as evidenced by Hellweg and Crisfield (1998). To address this limitation, the MRD method, as proposed by Chan (1988), was employed and proved highly effective. This method enabled the analysis to continue through the asymmetric bifurcation behavior, even when larger load increments were applied, demonstrating its superiority over traditional techniques.

2.9.2 Snap Through and Snap Back in Eight-Bar Trusses

The third problem examined an eight-bar truss structure, originally studied by Powell and Simons (1981) and shown in Fig. 2.10. The example too uses synthetic data, and the details are shown in Fig. 2.10. The structure exhibited snap-through behavior at node '7' and snap-back at node '1', and the results obtained in Fig. 2.11 are identical to those published by Powell and Simons, affirming the robustness of the current formulation in capturing complex postbuckling responses.

This problem also provided an opportunity to investigate the influence of second-order terms in Green's strain on the geometric nonlinear analysis of trusses. Following the work of Glabiz (1986), a comparison was made between results obtained with and without second-order strain terms. Figure 2.11 shows that including these terms significantly impacts the results, with the limit load in the linear case approximately

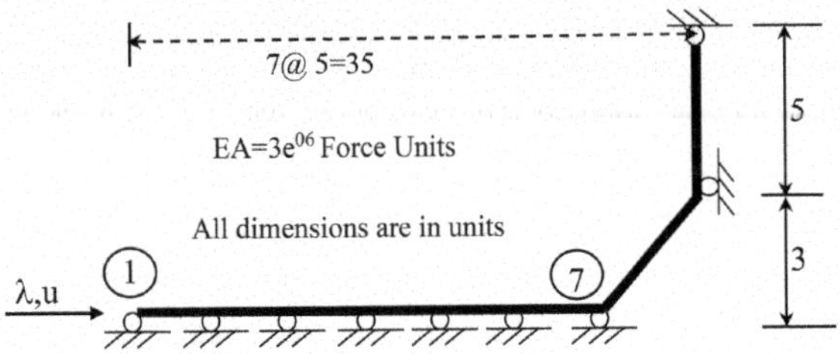

Fig. 2.10 Eight-bar truss exhibiting snap-through and snap-back behavior

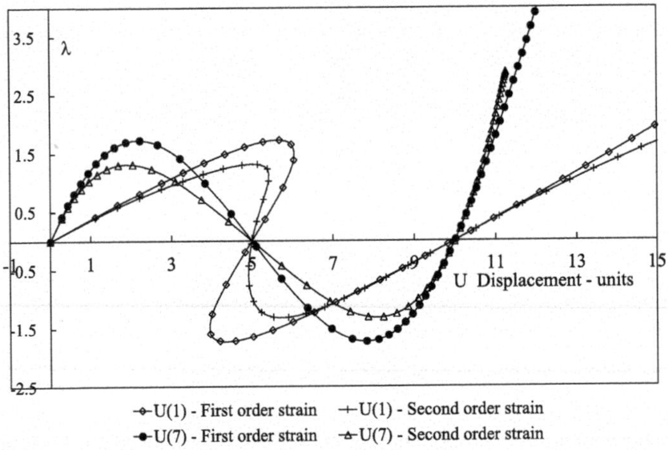

Fig. 2.11 Load deflection plot of eight-bar truss with order of terms—snap-through and snap-back behavior

18% higher than in the nonlinear case. This underscores the importance of including second-order strain terms when analyzing structures exhibiting limit behavior.

2.9.3 Postbuckling Analysis with Member Instability

The fourth and fifth problems dealt with truss structures exhibiting member instability during postbuckling analysis. The fourth problem (Fig. 2.12) is a 24-bar truss dome, where the buckling of the inner circle struts and the overall postbuckling behavior of the dome were analyzed. The results, shown in Fig. 2.13, demonstrate the efficiency of the CR-TL formulation in accurately capturing the postbuckling behavior. The findings align with those presented by Cichon (1984), reinforcing the effectiveness of the CR-TL formulation. The interaction of member instability and global instability is evident from this example.

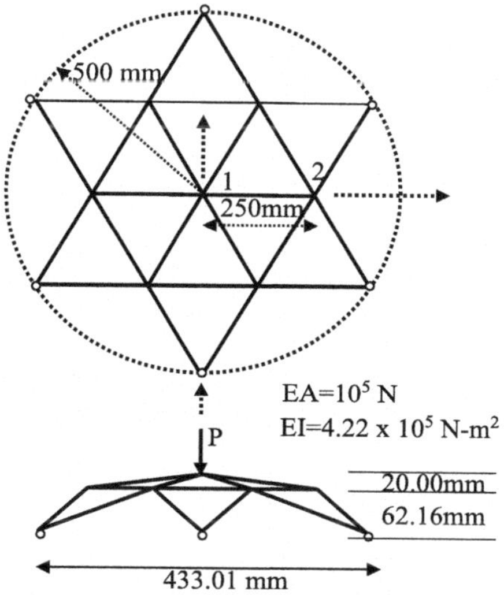

Fig. 2.12 Geometry of 24-bar truss dome

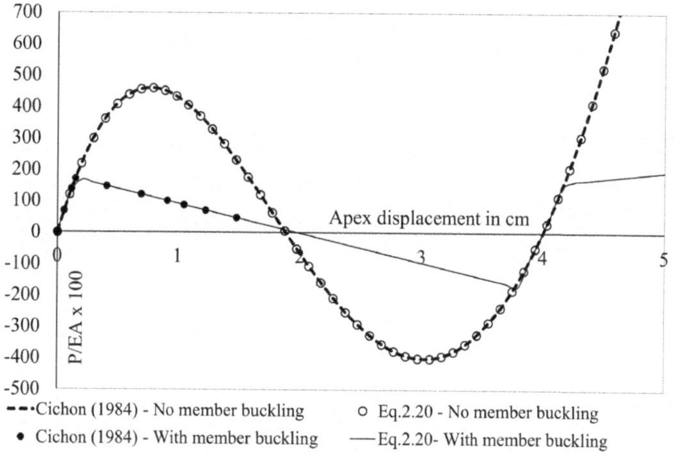

Fig. 2.13 Postbuckling response of 24-bar truss dome

2.9.4 Member Instability Demonstrated Using Thomson's Strut

The final example focused on the postbuckling analysis of Thomson's strut (Fig. 2.14), a structure with circular cross-sections. In this analysis, member instability was introduced in members 15 and 16, individually and in combination, by varying their cross-sectional properties as specified in Table 2.2.

Fig. 2.14 Thompson's strut under axial compression

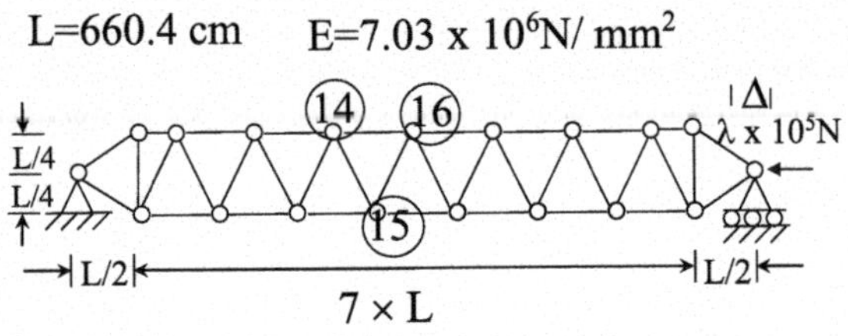

Table 2.2 Cross-sectional areas of Thomson's strut structure

Members	Area of cross-section		
	No member buckling and member 16 buckling (mm^2)	Member 15 buckling (mm^2)	Members 14 and 16 buckling (mm^2)
Top chord, bottom chord, and vertical members	5484	5484	5484
Diagonal members	5161	5161	5161
Member number 15	5484	5161	5484
Member numbers 14 and 16	5484	5484	5161

The postbuckling analysis results matched those reported by Kondoh and Atluri (1985), confirming the accuracy of the CR-TL approach. These findings are illustrated in Fig. 2.15, underscoring the importance of accounting for individual member buckling to accurately assess the global postbuckling behavior of truss systems.

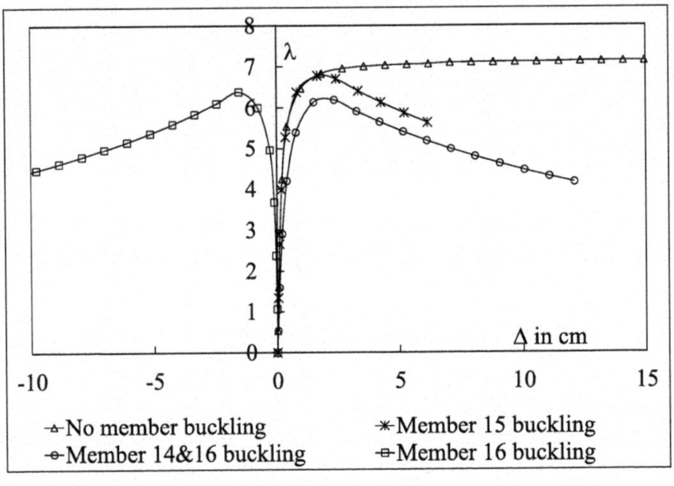

Fig. 2.15 Postbuckling paths of Thomson's strut

2.10 Summary

This chapter introduces a secant matrix method within a co-rotational finite element framework (CR-TL) for space truss analysis. The relationship between increments in co-rotated forces and natural force rotation was derived based on geometric principles. This formulation was integrated into a computational program and validated against several benchmark problems from the literature. The analysis showed that for trusses exhibiting postbuckling behavior, including second-order strain terms in Green's strain is critical for accuracy. Additionally, limitations in the conventional arc-length method were highlighted, particularly with sharp snap-back in postbuckling paths. To address this, the minimum residual displacement (MRD) method by Chan (1988) offered an effective solution, allowing the program to handle snap-back behavior efficiently. Finally, the importance of accounting for individual member instability was emphasized, with relevant equations provided for implementation.

References

Argyris JH, Balmer H, Doltsinis PC, Dunne M, Haase M, Kleiber M, Malejannakis GA, Mlejenek HP, Muller M, Scharpf DW (1976) Finite element method—the natural approach. Comput Meth Appl Mech Eng 17/18:1–106. https://doi.org/10.1016/0045-7825(79)90083-5

Belytschko T, Hseih BJ (1973) Nonlinear transient finite element analysis with convected coordinates. Int J Num Meth Eng 7:255–271. https://doi.org/10.1002/nme.1620070304

Bergan PG, Soreide TH (1973) A comparative study of different numerical solution techniques as applied to nonlinear structural problems. Comput Meth Appl Mech Eng 2:185–201. https://doi.org/10.1016/0045-7825(73)90014-5

Bergan PG, Soreide TH (1978) Solution of large displacement and instability problems using current stiffness parameter. In: Bergan PG (ed) Finite elements in nonlinear mechanics. Tapir, Trondheim

Blanford GE (1995) Large deformation analysis of inelastic space truss structures. J Struct Eng 122(2):407–415

Britvec SJ (1973) The stability of elastic systems. In: Pergamon Unified Engineering series. https://doi.org/10.1016/C2013-0-05352-2

Britvec SJ (1995) Stability and optimisation of flexible space structures. Birkhauser, Berlin. https://doi.org/10.1007/978-3-0348-9053-3

Chan SL (1988) Geometric and material nonlinear analysis of beam—columns and frames using the minimum residual displacement method. Int J Numer Meth Eng 26:2657–2669. https://doi.org/10.1002/nme.1620261206

Cichon C (1984) Stability analysis of elastic space trusses. Proceedings of Third International Conference on Space Structures. H.Nooshin (ed), Elsevier, 567–570

Crisfield MA (1991) Nonlinear finite element analysis of solids and structures—volume 1 Essentials. Wiley, New York, NY

Crisfield MA (1997) Nonlinear finite element analysis of solids and structures—volume 2 Advanced topics. Wiley, New York, NY

Glabiz (1986) The influence of second-order values in the numeric analysis of bar structures apt to snap through. In: Yagawa G, Atluri SN (eds) Computational Mechanics '86, Proceedings of International Conference on Computational Methods

Hangai Y, Kawamata S (1972) Nonlinear analysis of space frames and snap-through buckling of reticulated shell structures. In: IASS Symposium on Tension Structures and Space Frames, Architecture Institute of Japan

Hellweg HB and Crisfield MA (1998) A new arc-length method for handling sharp snap backs. Comput Struct 66(5), 705–709

Kondoh K, Atluri SN (1985) Influence of local buckling on global instability: simplified, large deformation postbuckling analysis of plane trusses. Comput Struct 21:613–627. https://doi.org/10.1016/0045-7949(85)90140-3

Kondoh K and Kazunavi Tanaka (1986) Large deformation and postbuckling analysis of plane and space trusses. Computational Mechanics '86, Proceedings of International Conference on Comput Methods G. Yagawa and S.N. Atluri (eds.), III 191–196

Madi UR (1984) Idealising the member behaviour in the analysis of pin-jointed spatial structures. In: Nooshin H (ed) Proceedings on Third International Conference on Space Structures. Elsevier

Mallet RH, Marcal PV (1968) Finite element analysis of nonlinear structures. J Struct Div 94:2081–2105. https://doi.org/10.1016/0045-7949(84)90173-1

Mattiasson K, Samuelsson A (1984) Total and updated Lagrangian forms of the co-rotational finite element formulations in geometric and material nonlinear analysis. In: Taylor et al (ed) Proceedings of International Conference on Numerical methods for nonlinear problems. Pineridge Press, Swansea

Meek JL (1971) Matrix structural analysis. McGraw Hill Book Company, NY, 526–588

Murtha Smith E (1994) Nonlinear analysis of space trusses. J Struct Eng 120(9):2717–2747

Narayanan G, Krishnamoorthy CS (1994) Investigations on truss element models for the static response of geometrically nonlinear structural systems. J Struct Eng 21:145–163

Noor AK (1974) Nonlinear analysis of space trusses. J Struct Eng 100:533–546. https://doi.org/10.1016/0045-7825(80)90027-4

Papadrakakis M (1987) Inelastic postbuckling analysis of space frames. In: Cheremisinoff PN, Cheng LS (eds) Civil engineering practice—structures. Technomic Publishing

Powell G, Simons J (1981) Improved iteration strategy for nonlinear structures. Int J Numer Meth Eng 17:1455–1467. https://doi.org/10.1002/NME.1620171003

Punniakotti NM, Liew JYR, Shanmugam NE (1995) PC based nonlinear analysis of space frames. In: Shanmugam NE, Choo YS (eds) Structural steel—PSSC-95, Proceedings of 4th Pacific Structural Steel Conference Vol. 1—Steel Structures. Pergamon

Rajasekaran S, Murray DW (1973) Incremental finite element matrices. J Struct Eng 99:2423–2438. https://doi.org/10.1061/JSDEAG.0003667

Tanaka K, Kondoh K, Atluri SN (1985) Instability analysis of space trusses using exact tangent stiffness matrices. Finite Elements in Analysis and Design, 1:291–311

Wempner G (1969) Finite elements, finite rotations and small strains of flexible shells. Int J Solids Struct 5:117–153. https://doi.org/10.1007/978-3-662-10326-5_28

McGuire W, Gallagher Richard H, Ziemian Ronald D (2015) Matrix Structural Analysis. Wiley, Hoboken, NJ

Yang YB, Leu LJ (1990) Postbuckling analysis of trusses with various formulations. AIAA J. 28:946–948. https://doi.org/10.2514/3.25146

Yang YB, Yang CT (1996) Postbuckling response of space trusses considering the effects of member buckling and yielding. In: Proceedings of Third Asian-Pacific Conference on Computational Mechanics

Zhu K, Al-Bermani FGA, Kitipornchai S (1993) Nonlinear dynamic analysis of lattice structures. Comput Struct 12:9–15. https://doi.org/10.1016/0045-7949(94)90250-X

Chapter 3
Variable Order Secant Matrices for the Kinematically Exact Postbuckling Analysis of Beams and Frames

Abstract This chapter presents a new technique called Variable Order Secant Matrix (VOSM) for the kinematically exact postbuckling analysis of structural components such as beams and frames. Using Reissner's strain–displacement relations, the VOSM approach uses higher-order secant matrices to obtain accurate results in the postbuckling behavior of plane beams, frames, and arches. The concept and the formulation of the VOSM are introduced using an example of an inextensible elastica. Kinematically exact Reissner's beam element is formulated and applied to standard benchmark problems to validate its capabilities. The numerical results illustrate the VOSM technique's effectiveness in representing postbuckling responses of plane beams, frames, and arches. This chapter also presents and validates many benchmark problems to help researchers and professionals.

Keywords Variable order secant matrix (VOSM) · Postbuckling analysis · Reissner's beam element · Large deformation · Structural nonlinearity

3.1 Nonlinear Analysis of Beams and Frames

Studying large deformation in beams and frames is important in structural analysis, particularly when studying complex postbuckling behavior. As structures like beams, frames, and arches experience large bending deformation, their strains may remain relatively small.

Various kinematic formulations have been developed to describe large deformation in beams. The three main approaches—Total Lagrangian (TL), Updated Lagrangian (UL), and co-rotational (CR)—differ in their treatment of reference configurations, as discussed in Chap. 2. Each formulation has its applications, but choosing the correct reference configuration is essential for analyzing framed structures that experience large deformations.

Early studies showed that the TL formulation did not capture large deformations and beam rotations, especially when using conventional quadratic secant matrix approaches. For example, the TL formulation proposed by Mallet and Marcal (1968),

A. Jayachandran, *Elastic Postbuckling Analysis of Structural Systems*, Solid Mechanics and Its Applications 232, https://doi.org/10.1007/978-3-031-89936-2_3

as well as those by Rajasekaran and Murray (1973) and Chajes and Churchill (1987), were shown to be insufficient for accurately describing the behavior of beams under large deflections. This was later attributed to the linearization of curvature in the TL approach, as identified by Mattiasson and Samuelsson (1984).

In response to the inadequacies of the TL formulation, many researchers used UL and CR formulations that were more suitable for large deflections. However, there has been renewed interest in using the TL framework to implement Reissner's (1972) kinematically exact beam formulation, which provides a more precise representation of beam kinematics.

This chapter introduces a novel approach called the Variable Order Secant Matrix (VOSM) technique to address the limitations of existing TL formulations. The VOSM technique enhances the accuracy of large deformation and postbuckling analysis for beams and frames by employing higher and variable order secant matrices.

A significant contribution of this chapter is the development of Reissner's kinematically exact beam element, a geometric nonlinear finite element formulation created to solve problems related to postbuckling behavior. This element has been applied to a range of benchmark problems involving beams, frames, and arches, providing valuable insights into their postbuckling behavior. These solved problems serve as a repository of benchmark studies, acting as a reference for future studies in the nonlinear analysis of beams, frames, and arches.

3.2 Notable Previous Research on Large Deflection Analysis of Beams and Frames

The analysis of large deformations in beams and frames has been a significant area of research. Early analytical methods were developed by Huddleston (1967, 1968), who presented solutions for the elastic large deformation of frames and arches. Antman (1968, 1972) provided finite deformation analytical solutions for hyperstatically loaded elastica of circular shapes, laying foundational work in this field.

The co-rotational (CR) formulation emerged as an effective approach for nonlinear beam analysis. Jennings (1968) introduced a co-rotational beam element for large deformation analysis of frames, followed by Belytschko and Hsieh (1973), who applied CR methods to beam nonlinear analysis. Argyris (1982) further advanced the CR formulation with his 'natural approach', significantly improving the modeling of large displacements and rotations in structures.

Simultaneously, Total Lagrangian (TL) and Updated Lagrangian (UL) formulations were explored. Epstein and Murray (1976) and Wood and Zienkiewicz (1977) presented TL formulations for large deformation analysis using higher-order strain–displacement relations. Oran and Kassimali (1976) developed an UL formulation employing a general beam-column method.

However, conventional quadratic TL formulations were reported to be inadequate for accurately capturing large deformation behaviors. Pignataro et al. (1982) highlighted the limitations of these models, noting their inability to account for curvature effects adequately. Boswell (1985) and Mattiasson et al. (1986) identified that approximations in curvature led to inaccuracies in second-order beam formulations, particularly in postbuckling analyses.

To overcome these challenges, researchers turned to kinematically exact beam theories. Reissner's (1972) kinematically exact beam formulation became very important for subsequent developments. Implementations of Reissner's theory were presented by Ibrahimbegovic and Frey (1993, 1995), who addressed membrane and shear locking issues using consistent linearization and reduced integration techniques. The combined effect of membrane and shear locking is alleviated using reduced integration. Hancock (1993) presented a formulation for the second-order elastic analysis of rigid jointed frames.

Khamlichi et al. (1998) carried out a comparative study of postbuckling beam formulation. They concluded that the linear critical load is least affected by the higher-order strain terms in the postbuckling range.

Pacoste and Erikkson (1995) presented a formulation for the post-critical analysis of frames. Work using finite element methods includes the work of Crisfield (1990), who developed a 3D beam element based on the natural approach, and Saje (1990, 1991), who formulated nonlinear beam models focusing on cross-sectional rotations. These formulations improved the accuracy of large deformation analyses without exhibiting locking phenomena.

Pi and Trahair (1996, 1998) emphasized the significance of higher-order terms in beam deformation. Their studies on the nonlinear analysis of arches demonstrated that higher-order curvature effects are crucial for accurate predictions, especially in deep arches. Efforts to enhance computational efficiency led to the concept of modeling one frame member with a single finite element. So and Chan (1991); Petrolito and Legge (1996); and Teh (2001) contributed to this area by developing higher-order beam elements capable of capturing complex behaviors with fewer elements.

Shear-locking issues in beam elements were addressed by Reddy (1997), who discussed methods to mitigate this problem in finite element formulations. Salerno and Lanzo (1997) and Saje et al. (1997) further refined kinematically exact beam theories, incorporating Koiter's perturbation analysis and addressing elastoplastic materials through the principle of virtual work.

Gummadi and Palazotto (1998) utilized Reissner's exact strain–displacement relations in beam analysis, effectively handling large rotations using curvilinear coordinates. Chakraborty and Mazumdar (1997) highlighted the necessity of retaining higher-order terms in TL formulations to prevent inadequate representation of rigid body modes in large deformation problems.

The literature includes Sami Pajunen's (1999, 2000) work, who presented finite element formulations based on Reissner's kinematically exact beam theory. Utilizing methods like Fried's orthogonal trajectory arc-length technique, these formulations effectively handled finite displacements and rotations in beam elements.

We build upon these developments in the present book by deriving kinematically exact beam formulations within the Total Lagrangian (TL) framework by addressing the shortcomings of conventional TL formulations. By introducing the Variable Order Secant Matrix (VOSM) technique, we aim to enhance the accuracy of large deformation and postbuckling analyses for beams, frames, and arches.

3.3 Variable Order Secant Matrix (VOSM) Method

The present section is built upon the fundamentals we learned in Sect. 1.6 of Chap. 1. The concept of variable order secant matrices (VOSM) is explained through a large deformation problem of a plane inextensible elastica. The Euler–Bernoulli hypothesis is assumed to hold; hence, the normal to the centroidal axis remains straight and normal throughout the deformation. The local beam axis 'x' is assumed to be inextensional during deformation. With the above assumptions, the exact curvature of the plane elastica in terms of the displacement gradients can be written as

$$\chi = \frac{\partial^2 w}{\partial x^2}\left(1 - \left(\frac{\partial w}{\partial x}\right)^2\right)^{-1/2}, \tag{3.1}$$

where 'w' is the assumed deflected shape of the beam. When 'E' is Young's modulus, and 'I' is the moment of inertia of the beam, the internal strain energy stored during large flexural deformation of the beam of length 'L' can be written as

$$U = \frac{EI}{2}\int_0^L \chi^2 dx = \frac{EI}{2}\int_0^L \xi dx. \tag{3.2}$$

Using Eq. (3.1), the integrand ξ in Eq. (3.2) can be expressed in terms of a binomial series as

$$\xi = \sum\left(\frac{\partial^2 w}{\partial x^2}\right)^2\left(\frac{\partial w}{\partial x}\right)^n \quad n = 0, 2, 4, 6, 8, \tag{3.3}$$

where 'n' represents the power of the displacement gradient $\partial w/\partial x$ or written as $w_{,x}$. This proves that the strains, curvature in this particular case, can have displacement gradients more than order two; in fact, it may have a generic value of 'n'. Hence, we can write Eq. (1.24) from Chap. 1 as

$$U(\mathbf{g}) = \frac{1}{2}\int\limits_V (U_2(\mathbf{g}) + U_3(\mathbf{g}) + U_4(\mathbf{g}) + U_5(\mathbf{g}) + U_6(\mathbf{g}) \ldots + U_{\mathrm{n}}(\mathbf{g}))dV. \tag{3.4}$$

Although Eq. (3.3) is an even function, Eq. (3.4) represents a generalized way of writing the strain energy using displacement gradients. It is seen that Eq. (3.4) can accommodate any order of monomials of $U(\mathbf{g})$. For the inextensible elastica considered $\mathbf{g}^{\mathrm{T}} = \{w_{,x}, w_{,xx}\}$.

In general, the displacement gradient vector $\{\mathbf{g}\}$ is given by, $\{\mathbf{g}\} = \{g_1, g_2, g_3, \ldots g_k\}^{\mathrm{T}}$, and $U_n(g_1, g_2, g_3, \ldots g_k)$ is a real-valued scalar function of degree equal to n, then U_n could be represented as:

$$U_n = \{\mathbf{g}\}^{\mathrm{T}}_{1\times k}\frac{1}{n(n-1)}\left[\nabla^2 U_n\right]_{k\times k}\{\mathbf{g}\}_{k\times 1}. \tag{3.5}$$

Therefore, if $U = U_2 + U_3 + U_4 + \ldots + U_n$, then

$$U = \mathbf{g}^{\mathrm{T}}\frac{1}{2}[\nabla^2 U_2]\mathbf{g} + \mathbf{g}^{\mathrm{T}}\frac{1}{6}[\nabla^2 U_3]\mathbf{g} + \mathbf{g}^{\mathrm{T}}\frac{1}{12}[\nabla^2 U_4]\mathbf{g} + \mathbf{g}^{\mathrm{T}}\frac{1}{20}[\nabla^2 U_5]\mathbf{g} + \ldots + \mathbf{g}^{\mathrm{T}}\frac{1}{n(n-1)}[\nabla^2 U_n]\mathbf{g}. \tag{3.6}$$

Equation (3.6) is rearranged as

$$U = \mathbf{g}^{\mathrm{T}}\left[\frac{1}{2}[\nabla^2 U_2] + \frac{1}{6}[\nabla^2 U_3] + \frac{1}{12}[\nabla^2 U_4] + \frac{1}{20}[\nabla^2 U_5] + \ldots. + \frac{1}{n(n-1)}[\nabla^2 U_n]\right]\mathbf{g}. \tag{3.7}$$

Note that $[\nabla^2 U_n]$ is a matrix of monomials with degree $(n-2)$. Thus, $[\nabla^2 U_2]$ is a constant matrix represented using $[\mathbf{K}]$, $[\nabla^2 U_3]$ is a matrix containing linear monomials in terms of the components of $\mathbf{g}$ represented using **N1**, $[\nabla^2 U_4]$ is a matrix containing quadratic monomials in terms of the components of $\mathbf{g}$ represented using **N2**, $[\nabla^2 U_5]$ is a matrix containing cubic monomials in terms of the components of $\mathbf{g}$ represented using **N3**, and so on. Thus,

$$U = \mathbf{g}^{\mathrm{T}}\left[\frac{1}{2}[\mathbf{K}] + \frac{1}{6}[\mathbf{N1}] + \frac{1}{12}[\mathbf{N2}] + \frac{1}{20}[\mathbf{N3}] + \ldots + \frac{1}{n(n-1)}[\mathbf{N(n-2)}]\right]\mathbf{g}. \tag{3.8}$$

For the inextensible elastica, assuming a suitable approximating polynomial for the beam's in-plane displacement 'w' and carrying out the summation in Eq. (3.3), the strain energy in Eq. (3.8) can be evaluated. If the nodal displacement vector $\mathbf{p} = \{v, \theta\}$ (where v is the deflection, and θ cross-sectional rotation), the displacement gradients can be related to nodal displacements through the relation

$$\mathbf{g} = \mathbf{D}\mathbf{p}. \tag{3.9}$$

After pre and post-multiplying with $\mathbf{D}$ and $\mathbf{D}^{\mathrm{T}}$ Eq. (3.8) is recast into a form that is a function of nodal displacement vector $\mathbf{p}$ as

$$U = \mathbf{p}^{\mathrm{T}}\left[\frac{1}{2}[\mathbf{K}] + \frac{1}{6}[\mathbf{N1}] + \frac{1}{12}[\mathbf{N2}] + \frac{1}{20}[\mathbf{N3}] + \ldots + \frac{1}{n(n-1)}[\mathbf{N(n-2)}]\right]\mathbf{p}. \tag{3.10}$$

For inextensible beam problems, the terms in Eq. (3.4), which are dependent on odd powers, vanish (Pignataro et al. 1991; Thomson and Hunt 1973), and Eq. (3.8) can be rewritten as

$$U = \mathbf{p}^{\mathrm{T}}\left[\frac{1}{2}[\mathbf{K}] + \frac{1}{12}[\mathbf{N2}] + \frac{1}{30}[\mathbf{N4}] \ldots + \frac{1}{n(n-1)}[\mathbf{N(n-2)}]\right]\mathbf{p}. \tag{3.11}$$

Now, total potential energy can be written assuming the corresponding generalized load vector as $\mathbf{q}$, therefore

$$\pi = U - \mathbf{p}^{\mathrm{T}}\mathbf{q}. \tag{3.12}$$

Applying the necessary condition of equilibrium $\frac{\partial \pi}{\partial \mathbf{p}} = 0$, and keeping in mind the order of dependence of term U, the equilibrium equation can be written as

$$\left[[\mathbf{K}] + \frac{1}{3}[\mathbf{N2}] + \frac{1}{5}[\mathbf{N4}] \ldots + \frac{1}{(n-1)}[\mathbf{N(n-2)}]\right]\mathbf{p} = \mathbf{q}. \tag{3.13}$$

Following the Mallet and Marcal (1968) nomenclature, the tangent stiffness matrices can be written by differentiating Eq. (3.13) once again with respect to $\mathbf{p}$ and neglecting higher-order powers as

$$\Delta\mathbf{q} = \mathbf{K}\Delta\mathbf{p} + \mathbf{N2}\Delta\mathbf{p} + \mathbf{N4}\Delta\mathbf{p} + \ldots \tag{3.14}$$

From a comparison of Eq. (3.13) and Eq. (3.14), the components matrices repeat themselves as per the Mallet and Marcal (1968) symbolism.

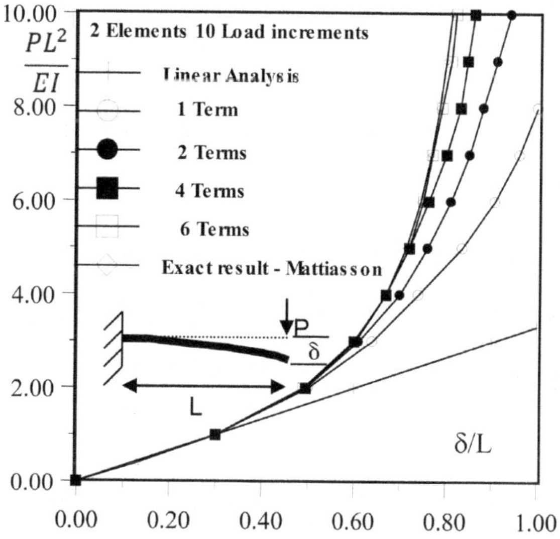

Fig. 3.1 Large deflections of a tip-loaded cantilever

3.4 Numerical Studies on Inextensible Elastica

The elastica is one of the elastic systems that can undergo very large displacement with small strains and is also a practical example to test the use of nonlinear finite elements, especially the VOSM method. Let us now vary the number of terms vis-à-vis the order of secant matrices in Eq. (3.13) and Eq. (3.14) in the solution of the large deformation of a tip-loaded cantilever. The beam is idealized with two elements. As seen from Fig. 3.1, at least six terms (up to secant matrix order 12), apart from the linear term, were needed to get exact solutions. As a further example, a large deflection of a cantilever subjected to a uniformly distributed load (UDL) was solved, and the results are shown in Fig. 3.2. It is seen from Fig. 3.2 that at least four terms (up to the secant matrix of order 8) in Eq. (3.13) and Eq. (3.14), in addition to the linear term, were needed to get a satisfactory solution. For beam elements with more than six terms, the results were as exact as the elliptic integral solutions of Mattiasson (1981). This is an example of how we can use the number of secant matrices depending upon the nonlinearity of the load-deformation behavior.

3.5 The Kinematically Exact Reissner Beam Element—VOSM Method

The kinematically exact theory of plane beams presented by Reissner (1972) has been presented in two different ways by Li (1997) and Pignataro et al. (1991). Consider a beam as a one-dimensional continuum in which the area of a cross-section is assumed

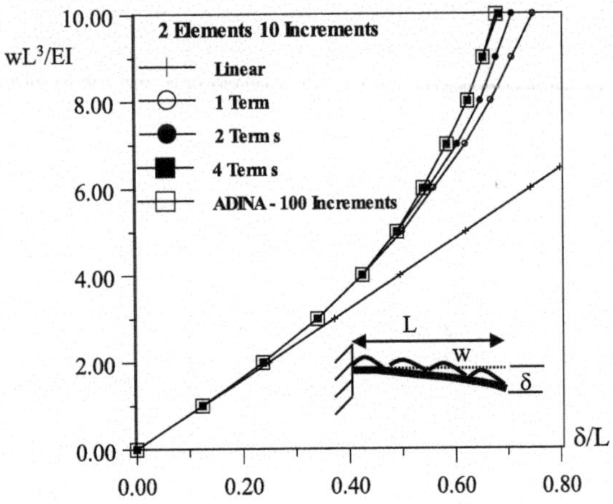

Fig. 3.2 Large deflection of a cantilever beam under UDL

to be concentrated along a regular curve called the beam axis, which is assumed to be rectilinear.

Each point along the beam axis is identified by $x \in [0, 1]$. By the general convention of representing beam deformation by two displacement components, namely, the rotation of the cross-section θ (Fig. 3.3) and translation $\mathbf{w} = u\mathbf{i} + v\mathbf{j}$. Adopting a Timoshenko beam theory, it is assumed that the cross-section, which was initially plane and normal to the neutral axis, will remain plane during deformation but, due to shearing, will not be perpendicular to the original beam axis.

Adopting a TL formulation, the unit vectors $\mathbf{a}_0$ and $\mathbf{b}_0$ are parallel to the Cartesian frame of reference $\mathbf{i}$, $\mathbf{j}$. The unit vectors $\mathbf{a}$, $\mathbf{b}$ in the deformed state can be related to the Cartesian frame of reference using orthogonality relation as

$$\mathbf{a} = \cos\theta\mathbf{i} + \sin\theta\mathbf{j}, \tag{3.15}$$

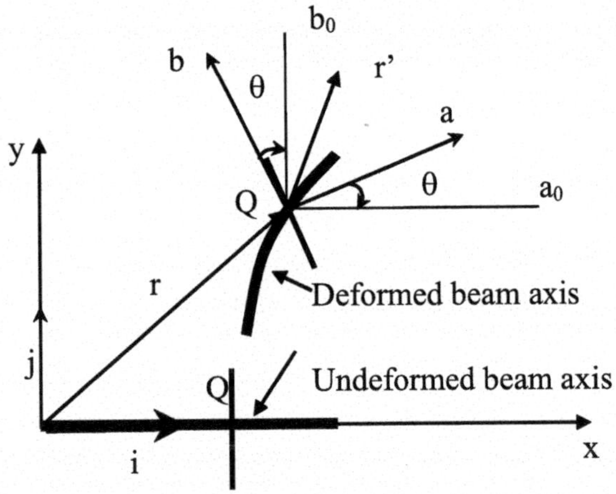

Fig. 3.3 Undeformed and deformed beam axis

$$\mathbf{b} = -\sin\theta\mathbf{i} + \cos\theta\mathbf{j}. \tag{3.16}$$

Now, the position vector '**r**' of the point '*Q*' at the deformed state can be represented as

$$\mathbf{r} = x\mathbf{i} + \mathbf{w}. \tag{3.17}$$

Using the deformation measures $\varepsilon, \gamma, \chi$ defined by Antman (1968, 1972), the vector $\mathbf{r}'$ could written as

$$\mathbf{r}\prime = (1+\varepsilon)\mathbf{a} + \gamma\mathbf{b}, \tag{3.18}$$

$$\chi = \theta_{,x}.$$

Here, $\varepsilon, \gamma, \chi$ represent the axial, shear, and flexural strains, respectively. By using the convention where differentiation is denoted by commas between preceding and succeeding indices and by incorporating Eq. (3.17) and Eq. (3.18), the Green–Lagrange strain for exact deformation is expressed as follows:

$$\begin{aligned}
\varepsilon &= \left(1+u_{,x}\right)\cos\theta + v_{,x}\sin\theta - 1,\\
\gamma &= v_{,x}\cos\theta - \left(1+u_{,x}\right)\sin\theta,\\
\chi &= \theta_{,x}.
\end{aligned} \tag{3.19}$$

These strain–displacement relations are exact as they eliminated the rigid body displacements, as formulated by Reissner (1972). Assuming a linear constitutive relationship, we define the second Piola–Kirchhoff internal forces as follows:

$$\begin{aligned}
N &= \int_A E\,\varepsilon\, dA,\\
Q &= \int_A G\,\gamma\, dA,\\
M &= \int_A y\,E\,\varepsilon\, dA,
\end{aligned} \tag{3.20}$$

where E and G are the Young's and shear moduli, respectively, A represents the cross-sectional area, and y denotes the distance from the centroidal axis. Using Eq. (3.20), the strain energy U_e for finite deformation can be expressed as:

$$U_e = \frac{1}{2}\int_o^1 (EA\varepsilon^2 + GA_S\gamma^2 + EI\chi^2)dx, \tag{3.21}$$

where A_s is the adjusted shear area. Substituting the strain expressions from Eq. (3.19) into Eq. (3.21), we obtain:

$$U_e = \frac{1}{2}\int_{o}^{1} \left[EA\{\left(1 + u_{,x}\right)\cos\theta + v_{,x}\sin\theta - 1\}^2 \right.$$
$$\left. + GA_s\{v_{,x}\cos\theta - \left(1 + u_{,x}\right)\sin\theta\}^2 + EI\theta_{,x}^2\right]dx. \tag{3.22}$$

Upon expansion, Eq. (3.22) results in sixteen terms in the integrand. To facilitate VOSM method implementation, the trigonometric terms are expressed in series form:

$$U_e = \frac{1}{2}\int_{o}^{1} (EA(\cos^2\theta + u_{,x}^2\cos^2\theta + 2u_{,x}\cos^2\theta + v_{,x}^2\sin^2\theta{+}1$$
$$+v_{,x}^2\sin 2\theta + u_{,x}v_{,x}\sin 2\theta - 2v_{,x}\sin\theta - 2\cos\theta$$
$$- 2u_{,x}\cos\theta) + GA_s(v_{,x}^2\cos^2\theta + \sin^2\theta + u_{,x}^2\sin^2\theta$$
$$+ 2u_{,x}\sin^2\theta - v_{,x}\sin 2\theta + u_{,x}v_{,x}\sin 2\theta) + EI\theta_{,x}^2)dx, \tag{3.23}$$

$$\cos = \frac{(-1)^n\theta^{2n}}{2n!}, n = 0, 1, 2, 3, 4\ldots, \tag{3.24}$$

$$\sin\theta = \frac{(-1)^{n-1}\theta^{2n-1}}{(2n-1)!}, n = 1, 2, 3, 4\ldots \tag{3.25}$$

For simplification, let $\alpha = 2\theta$, yielding the following series representations:

$$\cos^2\theta = \frac{(-1)^n\alpha^{2n}}{2(2n!)}, n = 0, 1, 2, 3, 4\ldots, \tag{3.26}$$

$$\sin^2\theta = \frac{(-1)^{n-1}\alpha^{2n}}{2(2n!)}, n = 1, 2, 3, 4\ldots, \tag{3.27}$$

$$\sin 2\theta = \frac{(-1)^{n-1}\alpha^{2n-1}}{(2n-1)!}, n = 1, 2, 3, 4\ldots \tag{3.28}$$

Applying Eqs. (3.24) through (3.28) into Eq. (3.23), the expression for strain energy U_e during large deformations can be written as

$$U_e = \frac{1}{2}\int_{o}^{1} \left(EA(\frac{(-1)^n\alpha^{2n}}{2(2n!)} + \frac{u_{,x}\alpha^{2n}}{2(2n!)} + \frac{2u_{,x}(-1)^n\alpha^{2n}}{2(2n!)} + \frac{v_{,x}^2(-1)^{n+1}\alpha^{2n}}{2(2n!)}\right.$$
$$+ \frac{v_{,x}(-1)^{n-1}\alpha^{2n-1}}{(2n-1)!} + \frac{u_{,x}v_{,x}(-1)^{n-1}\alpha^{2n-1}}{(2n-1)} - \frac{2v_{,x}(-1)^n\,\alpha^{2n-1}}{(2n-1)!}$$

$$
\begin{aligned}
&-\frac{2(-1)^{n-1}\theta^{2n}}{2n!}+\frac{2u_{,x}(-1)^{n}\alpha^{2n}}{2(2n!)})+\mathrm{GAs}(\frac{u_{,x}^{2}(-1)^{n}\alpha^{2n}}{2(2n!)} \\
&+\frac{(-1)^{n+1}\alpha^{2n}}{2(2n!)}+\frac{u_{,x}(-1)^{n+1}\alpha^{2n}}{2(2n!)}+\frac{u_{,x}(-1)^{n+1}\alpha^{2n}}{2(2n!)} \\
&+\frac{v_{,x}\,(-1)^{n}\alpha^{2n-1}}{2(2n!)}+\frac{u_{,x}v_{,x}(-1)^{n}\alpha^{2n}}{2(2-n)!})+EI\theta_{,x}^{2}\Bigg)dx.
\end{aligned}
\tag{3.29}
$$

Now define a gradient vector $\mathbf{g}^{\mathrm{T}} = \{\mathbf{u}_x, \mathbf{v}_x, \boldsymbol{\theta}\}$, representing displacement gradients within the strain–displacement relation (Eq. (3.19)). As per the formalism of Mallet and Marcel (1968), the strain energy is written in quadratic form, suitable for finite element integration as

$$U_e = \mathbf{g}_{1\times 3}^{\mathrm{T}}\overline{\mathrm{U}}_{3\times 3}\mathbf{g}_{3\times 1}. \tag{3.30}$$

Detailed derivation of the $\overline{\mathrm{U}}$ matrices are omitted here but presented explicitly in Appendix. Considering $\boldsymbol{\theta}$ as an independent variable, we employ an isoparametric three-noded beam element, with nodal degrees of freedom $\mathbf{p} = \{u, v, \theta\}$ (where u is axial deformation, v is deflection, and θ is the cross-sectional rotation). Displacement gradients are related to nodal displacements through:

$$\mathbf{g} = \mathbf{D}\mathbf{p}, \tag{3.31}$$

where $\mathbf{D}$ is a transformation matrix linking displacement gradients to nodal displacements. Substituting Eqs. (3.30) and (3.31) into Eq. (3.29) and integrating, we express strain energy U as (the details of the integration are explained in Chaps. 1 and 2)

$$U = \mathbf{p}^{\mathrm{T}}\left[\frac{1}{2}[\mathbf{K}] + \frac{1}{12}[\mathbf{N2}] + \frac{1}{30}[\mathbf{N4}]\ldots + \frac{1}{n(n-1)}[\mathbf{N(n-2)}]\right]\mathbf{p}. \tag{3.32}$$

In this section $\mathbf{p}$ represents the generalized displacement vector of an isoparametric beam element, depicted in Fig. 3.4. Equation (3.32) or Eq. (3.33) applies to any geometrically nonlinear beam problem. By substituting Eq. (3.31) and performing the integration, Eq. (3.29) can be expanded in series form, as each term is representable in a series. Recognizing that higher-order derivatives of U depend on terms of the form $\{\mathbf{p}\}^{n+2}$, we use the notation of Mallet and Marcal (1968) to express them in terms of $\mathbf{p}^n$ as follows:

In Eq. (3.34 or 3.33), the subscript indicates the order of dependence on the displacement vector $\mathbf{p}$. This quadratic form is especially well-suited for finite element implementation. Adding the potential energy contributions from external loads, the total potential energy Π is given by

$$\pi = \mathbf{p}^{\mathrm{T}}\left[\frac{1}{2}[\mathbf{K}] + \frac{1}{12}[\mathbf{N2}] + \frac{1}{30}[\mathbf{N4}]\ldots + \frac{1}{n(n-1)}[\mathbf{N(n-2)}]\right]\mathbf{p} - \mathbf{p}^{\mathrm{T}}\mathbf{q}, \tag{3.33}$$

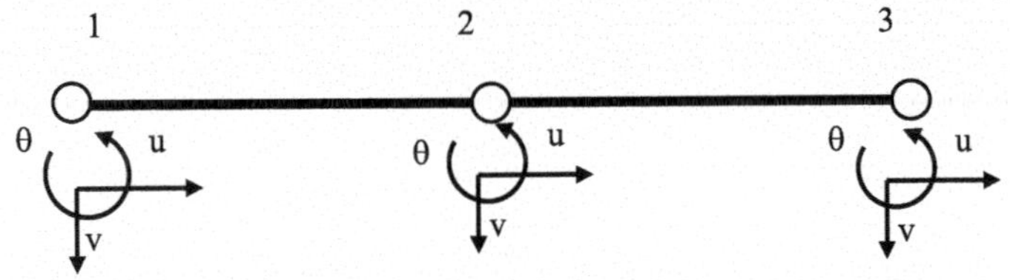

Fig. 3.4 Three-noded isoparametric beam element

where **q** represents the external load vector. By applying the equilibrium condition $\frac{\partial \pi}{\partial p} = 0$, and considering the order of each term, the equilibrium equation can be symbolically expressed as:

$$\mathbf{Kp} + \frac{1}{2}\mathbf{N}_1\mathbf{p} + \frac{1}{3}\mathbf{N}_2\mathbf{p} + \frac{1}{4}\mathbf{N}_3\mathbf{p} + \frac{1}{5}\mathbf{N}_4\mathbf{p} + \ldots = \mathbf{q}. \qquad (3.34)$$

K is the linear stiffness matrix, and **N1, N2, N3, N4, N5**… represent the higher-order secant matrices. Equation (3.34) defines the secant matrix relation, facilitating equilibrium assessment during Newton iterations. By expanding Eq. (3.34) as a Taylor series and neglecting higher-order terms, we obtain the tangent stiffness relationship:

$$\Delta\mathbf{q} = \mathbf{K}\Delta\mathbf{p} + \mathbf{N1}\Delta\mathbf{p} + \mathbf{N2}\Delta\mathbf{p} + \mathbf{N3}\Delta\mathbf{p} + \mathbf{N4}\Delta\mathbf{p} + \ldots \qquad (3.35)$$

As previously discussed, the variable-order secant matrices repeat themselves in energy, secant, and tangent stiffness formulations. These matrices can be determined by adjusting scalar factors, as indicated in Eqs. (3.33), (3.34) and (3.35). The explicit VOSM expressions are given in Appendix. Since these matrices are functions of expandable series, they can be generated for any desired order in a computational routine using a do-loop that specifies the matrix order in Eqs. (3.34) and (3.35).

3.6 Applications of the VOSM Method

The variable order secant matrices for the kinematically exact large deformation analysis of Reissner–Mindlin developed in this study have been tested against many benchmark problems published in the literature. The standard large deformation of a cantilever problem was chosen as the first example. The results of the problem for various order of terms and thus various orders of secant matrices in Eq. (3.34) and Eq. (3.35) are shown in Fig. 3.5. It is seen from Fig. 3.5 that when four secant matrices (terms N1 to N4) were used, the results were completely erroneous as compared to the exact elliptic integral solution of Mattiasson (1981). As the number of terms in the secant matrices increased, the results tended to become closer to the exact solution[2].

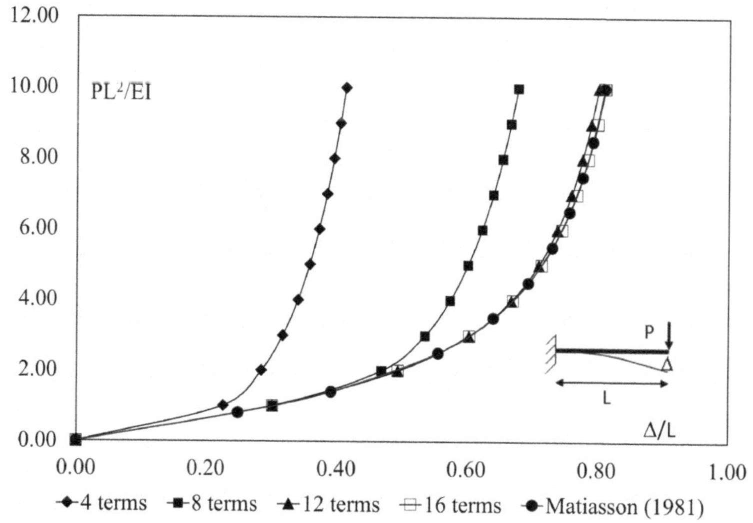

Fig. 3.5 Convergence study of tip-loaded cantilever

A seen from Fig. 3.5, when the number of incremental matrices was increased beyond sixteen, the present results coincided exactly with the elliptic integral solutions of Mattiason (1981). As stated earlier, the various secant matrices were formed using a simple do-loop in the computer program. As identified in the review of literature, the 'one element per member' concept has been the subject of research by many researchers (Saje 1991; Chan and Zhou 2000). Several commercial FE codes usually require breaking up one frame member into several beam elements to get reasonably accurate solutions (Ramesh and Krishnamoorthy 1993). In the present study, a frame member can be modeled using a single beam element if more than sixteen secant matrices are considered.

For the formulation presented in this chapter, there is no increase in the problem size (number of degrees of freedom) compared with the quadratic beam element. A parametric study found that the time taken for forming the extra higher-order secant matrices is negligible compared to the solution time.

The same example of large deflection of the cantilever was solved using one beam element and one increment of loading. The results are plotted in Fig. 3.6. The results show good correlation with the exact elliptic integral solutions of Mattiasson (1981).

The numerical values of the solution of the large deflection cantilever problem for three stages of loading $PL^2/EI = 4.0, 6.0, 10.0$ are compared with the results of leading researchers. Table 3.1 compares axial and in-plane displacement of the large deflection of the cantilever with the number of elements and loading increments. A comparison shows that the results of Saje (1991) are excellent in terms of the number of elements, loading increments and accuracy. Nevertheless, the formulation of Saje (1991) requires a single beam element to be divided into several parts having nodes, and due to the increase in the nodes, the order of the Gauss integration procedure

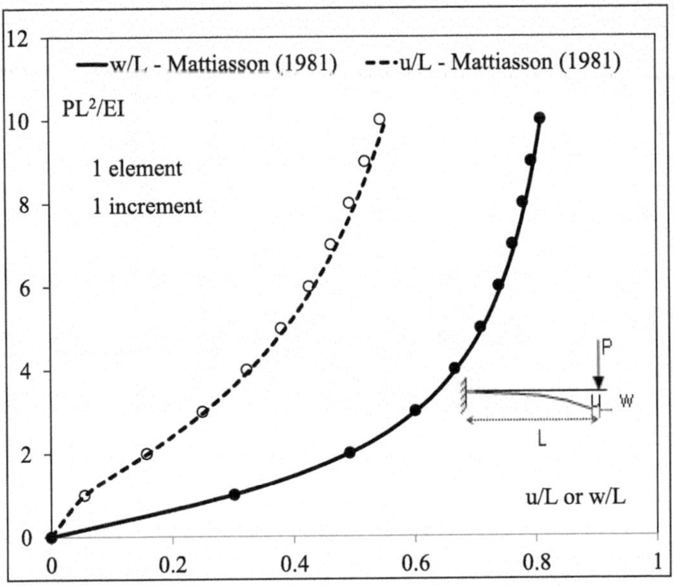

Fig. 3.6 Large deflections of a cantilever with one element and one increment using variable order secant matrix (VOSM) technique

has to be increased. It is seen that the results of the present chapter were as accurate as the elliptic integral solutions with four elements and five loading increments. The results of Morteza et al. (1997), who formulated the large deformation finite element equations with quadratic strain–displacement relations, are also presented in Table 3.1. Good results could not be obtained even with 20 elements and 500 increments. This illustrates the inadequacy of the quadratically formulated beam problems for large deformations, as described by Mattiasson and Samuelsson (1984).

Another benchmark problem suggested in the literature is the closed-form solutions for large deformation bending of a cantilever beam with two concentrated loads analytically solved by Frisch-Fay (1961). The problem is shown in Fig. 3.7, and the comparison of results of the present study with various researchers is shown in Table 3.2. With a discretization of two elements and one increment, the results of the present study compare well with the results of the Frisch-Fay (1961).

The next problem chosen to illustrate the performance of the present formulation is a cantilever beam subjected to an end moment. As seen from Fig. 3.8, at a moment value of $M_0 = 1.0$, the cantilever is completely bent into a circle. The slope of the cantilever at support must be zero. The results of the present study showed an error in the computation of slope less than 0.1% at support. Twenty-four secant matrices were needed to get good results for this problem.

The next benchmark problem is the postbuckling of Timoshenko cantilever beam with a small disturbing end moment. The shear parameter is defined using $\mu = E/kG$ (E—Young's modulus, k—shear correction parameter, and G is the shear modulus).

Table 3.1 Comparison of the results of the large deflection analysis of a cantilever with other researchers

Author	Formulation	No. of elements	No. of increments	PL^2/EI	u/L	w/L
Eiki et al. (1999)	Degenerate formulation	10	–	4.0	–	0.669
Equation (3.34) and Eq. (3.35)	VOSM	4	5	4.0	0.328	0.669
Mattiasson (1981) Elliptic integral solution				4.0	0.328	0.669
Gummadi and Palazotto (1998)	TL	10		6.0	–	0.731
Equation (3.34) and Eq. (3.35)	VOSM	4	5	6.0	0.434	0.744
Mattiasson (1981) Elliptic integral solution				6.0	0.434	0.744
Morteza et al. (1997)	TL	20	500	10.0	0.756	0.855
Ibrahimbegovic (1993)	TL	15	2	10.0	0.554	0.810
Saje (1991)	TL	1*	1	10.0	0.555	0.810
Equation (3.34) and Eq. (3.35)	VOSM	4	5	10.0	0.555	0.810
Mattiasson (1981) Elliptic integral solution				10.0	0.555	0.810

*7 Seven node element, 8 Gauss points

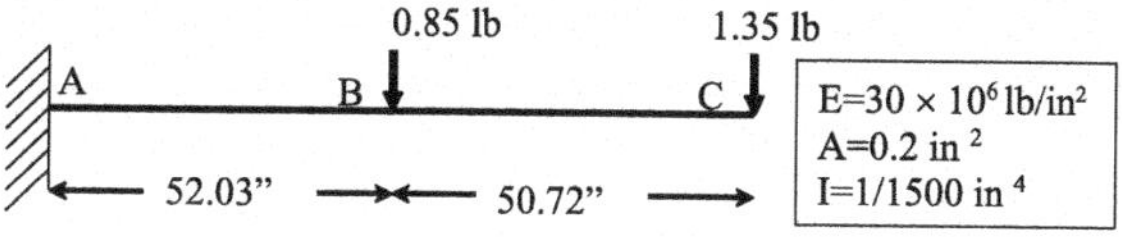

Fig. 3.7 Cantilever problem analyzed by Frisch-Fay (1961)

The problem of extensible and shear deformable elastica was solved by Goto et al. (1990) using elliptic integral solutions for three values of $\mu = 0, 3, 10$.

The results of the present study produced almost the same results as that of Goto et al. (1987) as shown in Fig. 3.9.

From the fundamental behavior of columns we know that an ideal slender column loses its stability after attaining the Euler critical stress. Nevertheless, an imperfect column continues to deflect from the inception of loading laterally and starts losing its stability near the Euler critical load.

Modern codes of practice (British code BS:5950 Part-I, Australian code AS:4100 1998) require second-order analysis of multistory frames using a notional load approach, calling for the second-order analysis of predominantly gravity-loaded sway steel frames subjected to a small lateral load. Such frames are susceptible to P-Δ effects. In such cases, the numerical solution is very sensitive in the vicinity of the critical load, and many commercially available programs produce erroneous

Table 3.2 Comparison of results of a cantilever analyzed by Frisch-Fay (1961)

Author	Formulation	No. of Elem	No. of increments	Horizontal deflection. at B	Vertical deflection at B	Horizontal deflection at C	Vertical def. at C (in.)
				Inches			
Jennings (1968)	CR-TL	20	100	4.94	20.60	15.01	51.14
Oran and Kassimali (1976)	Eulerian	2	20	7.89	25.26	30.17	68.88
Hsiao and Hou (1987)	CR	2	2	–	–	31.31	67.23
Li (1997)	TL	16	10	–	–	30.79	67.03
Equation (3.34) and Eq. (3.35)	VOSM	2	1	8.19	24.96	30.89	67.17
Exact value Frisch-Fay (1961)		–	–	8.28	25.14	31.01	67.32

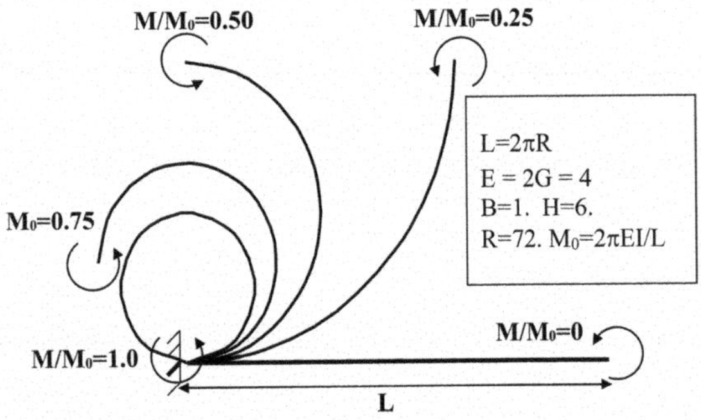

Fig. 3.8 Large deflections of a flexible cantilever subjected to end moment

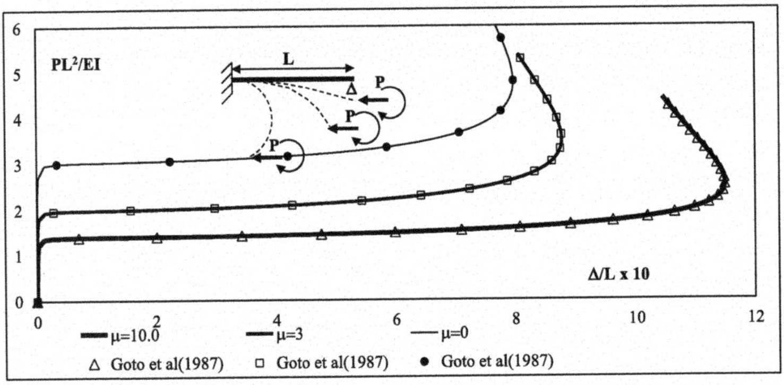

Fig. 3.9 Effect of shear deformation on the postbuckling of a cantilever column

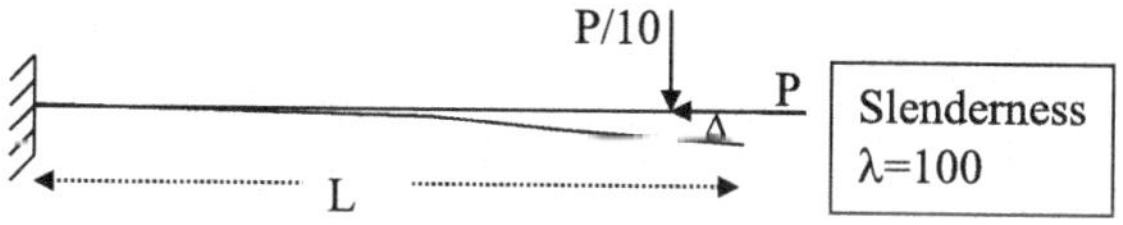

Fig. 3.10 Beam under compression and shear (benchmark problem of Australian institute of steel construction)

results. To assess the computer codes used to solve such problems, the Australian Institute of Steel Construction (AISC) has published several benchmark problems. The results of the present study passed through all the benchmark problems suggested by Petrolito and Legge (1995) and Duxbury et al. (1989). Two such examples are presented in this chapter. The first problem is a cantilever subjected to axial compression and a small lateral (shear) load. The problem is shown in Fig. 3.10. The results of the present study for one element per member and two elements per member are shown in Table 3.3. The numerical values agree very well with the results of AISC authored by Petrolito and Legge (1995).

The second benchmark frame of the Australian Institute of Steel Construction is a two-story portal frame subjected to a predominant gravity load and a small lateral load (Fig. 3.11). A comparison of the results of the present study with that of Petrolito and Legge (1996) is presented in Table 3.4. The sway deflection and moments results compare extremely well with the benchmark results, even with one element per frame member.

Roorda's frame is chosen as the next example for comparison, which undergoes asymmetric postbuckling. From the results shown in Fig. 3.12 for positive and negative values of ε, the frame exhibits a stable and unstable postbuckling behavior. The indirect verification of the result of the present study shows that the tangent line drawn between the stable and unstable postbuckling paths passes through the point $\lambda/\lambda_b = 1.0$.

Table 3.3 Comparison of benchmark result of Australian institute of steel construction—beam under compression and shear

	Present one element/ member one increment	Present two element/ member one increment	Petrolito and Legge (1996)	Present one element/ member one increment	Present two element/ member one increment	Petrolito and Legge (1996)
PL^2/EI	Tip deflection Δ/L			Moment at fixed end ML/EI		
0.5	0.020	0.020	0.020	0.060	0.060	0.060
1.0	0.055	0.055	0.055	0.154	0.155	0.155
1.5	0.121	0.123	0.123	0.330	0.333	0.333
2.0	0.279	0.285	0.285	0.741	0.760	0.760
2.25	0.403	0.419	0.419	0.110	0.114	0.114

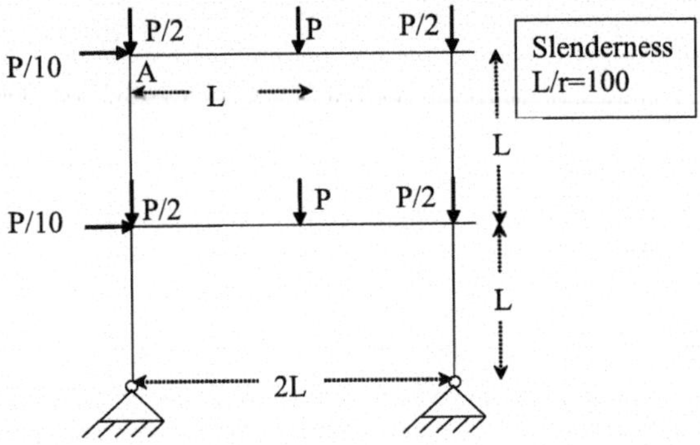

Fig. 3.11 Two-story frame under compression and sway loads (benchmark problem of Australian institute of steel construction)

Table 3.4 Comparison of benchmark results of Australian institute of steel construction story frame under compression and sway load

	Present one element/ member	Present two element/ member	Petrolito and Legge (1996)	Present one element/ member	Present two element/ member	Petrolito and Legge (1996)
PL^2/EI	Sway deflection u/L at 'A'			Moment ML/EI at 'A'		
0.1	0.011	0.011	0.011	0.015	0.015	0.015
0.2	0.026	0.026	0.026	0.030	0.030	0.030
0.3	0.047	0.047	0.047	0.044	0.044	0.044
0.4	0.080	0.080	0.080	0.054	0.053	0.033
0.5	0.136	0.137	0.137	0.056	0.056	0.056
0.6	0.258	0.256	0.257	0.041	0.040	0.040

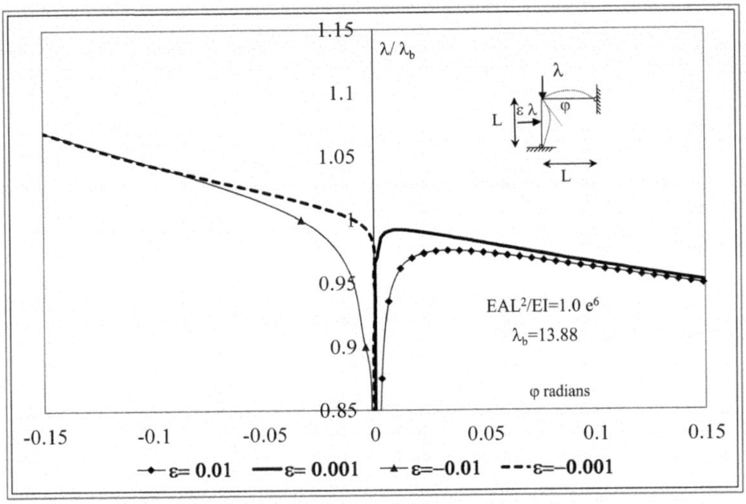

Fig. 3.12 Asymmetric postbuckling of Roorda's frame

Fig. 3.13 Diamond frame for rigid body motion test

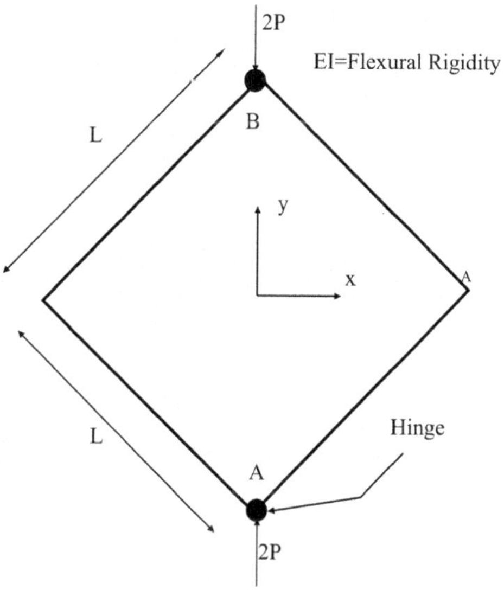

The next problem is a rigid body motion test problem suggested by Yang and Chiou (1986): a diametrically compressed diamond-shaped frame, as shown in Fig. 3.13.

The results of the present study are compared with the elliptic integral solutions of Matiasson (1981) in Table 3.5. It can be seen from Table 3.5 that the present formulation passes the rigid body motion test suggested by Yang and Chiou (1986).

Figure 3.14 presents the elastica deformation of Kerr's square frame (shown in Fig. 3.13 subjected to compression for various values of PL^2/EI. Each side of the frame is discretized using two finite elements. It is seen from Fig. 3.14 that the present formulation can model large deformation and large rotation very efficiently.

The next example problem is the famous Lee's frame problem (Fig. 3.15), which exhibits a snap-back behavior in the degree of freedom Δ. Results from Eq. (3.34) and Eq. (3.35) and the 'current stiffness parameter' (Bergan and Soreide 1973) are plotted. The results of the present study agree very well with the results of Shakurzadeh et al. (1999). From the plot of the current stiffness parameter (CSP) in Fig. 3.16, it is seen that the CSP is constant at a value of one when the load–deflection behavior is linear. The slope of the CSP plot represents the degree of nonlinearity of the problem. A

Table 3.5 Comparison of deformation of the diamond frame

Values at $PL^2/EI = 10$				
No. of elements	u_A/L	v_B/L	θ_B	M_AL/EI
2	−0.122	1.295	1.336	−5.841
4	−0.127	1.305	1.342	−5.800
Exact (Matiasson 1981)	−0.127	1.305	1.342	−5.798

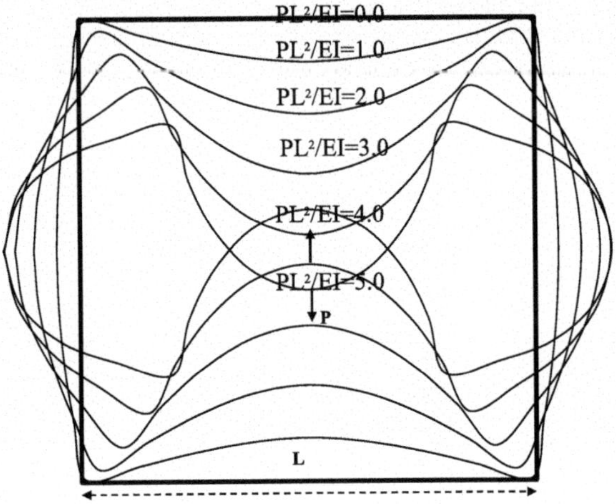

Fig. 3.14 Deformed shapes of Kerr's square frame

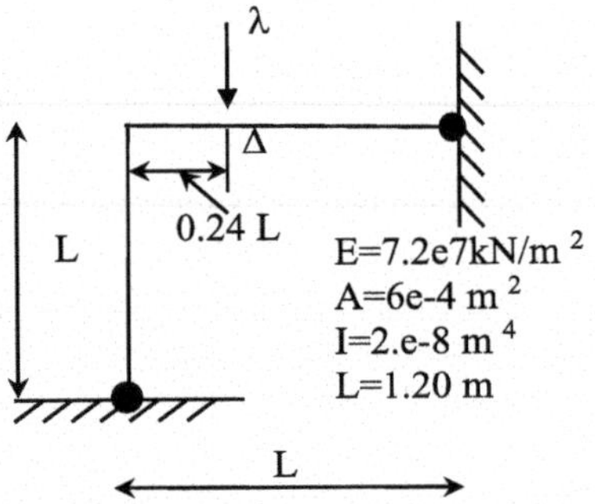

Fig. 3.15 Lee's frame, which undergoes snap-back buckling

positive slope of the CSP plot refers to a stiffening system, and a negative slope of the CSP plot represents a softening system.

The next frame problem is the symmetric sway buckling of a rectangular frame (Oran and Kasimali 1976). The frame, which is subjected to gravity load, is acted upon by a small transverse force, very similar to the application of notional loads. The postbuckling behavior of frames for two values of lateral disturbing force is presented in Fig. 3.17. It is seen from Fig. 3.17 that for $\varepsilon = 0.0001$ the frame exhibits a near bifurcational behavior. The load-deformation curve is almost flat after bifurcation. The 'minimum residual method' has been used to solve this problem and no numerical difficulties were encountered in traversing this postbuckling path.

To further validate the proposed formulation, two arch problems were analyzed. The first case, a circular arch, originally introduced by Da Deppo and Schmidt (1975), has been studied extensively. The arch model was discretized into 8, 10, and 16 elements to assess the accuracy of the formulation. Figure 3.18 displays the deformed shapes of the arch under a concentrated load applied at various levels, with material and geometric properties outlined in Fig. 3.19. Vertical and horizontal displacements

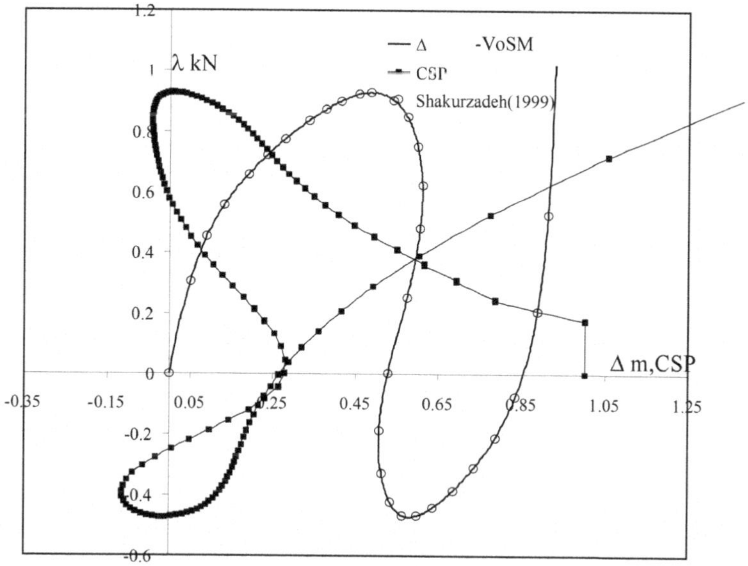

Fig. 3.16 Postbuckling behavior of Lee's frame

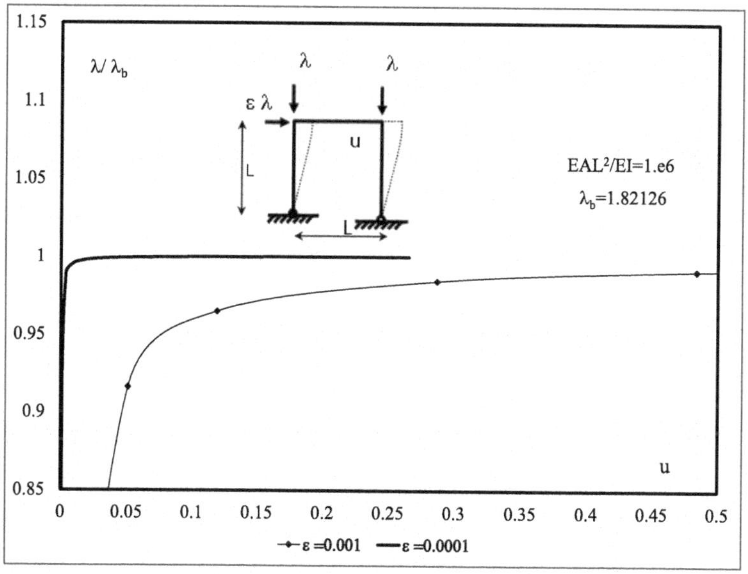

Fig. 3.17 Symmetric sway buckling of a square frame

of the arch crown in the postbuckling regime are also shown in Fig. 3.19. Comparative results, provided in Table 3.6, demonstrate the agreement between this study's outcomes and those in the existing literature.

Table 3.6 shows the results for the circular arch problem, reported by Ibrahimbegovic and Frey (1993), which was achieved using a discretization of 20 elements.

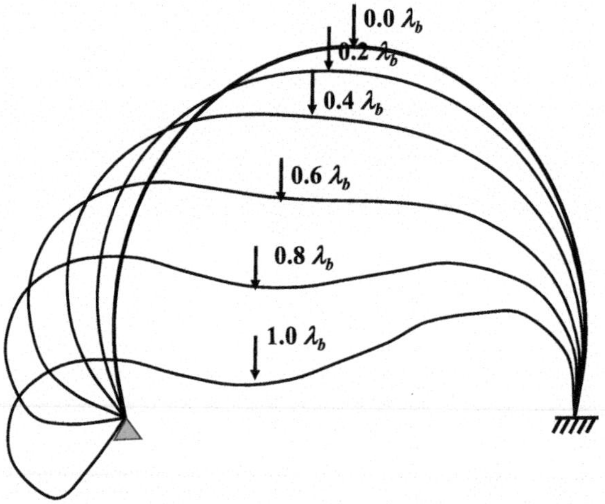

Fig. 3.18 Circular arch geometry and deformed shape

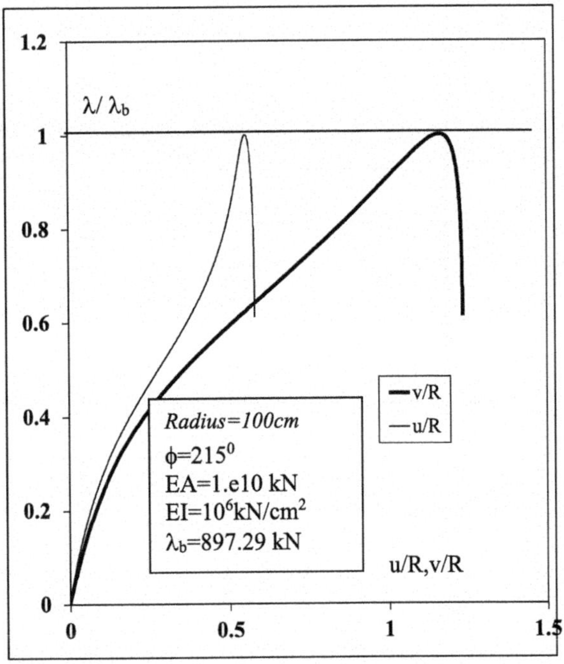

Fig. 3.19 Postbuckling behavior of circular arch

Table 3.6 Comparison of critical load of the circular arch

Authors	No. of elements	Critical load	% error
Dadeppo and Schmidt (1975)	200 (FD points)	897.0	0.03
Wood and Zienkiewicz (1977)	16	924.0	2.98
Hsiao and Hou (1987)	30	938.0	4.54
Ibrahimbegovic and Frey (1993)	20	897.5	0.02
Borri and Bottasso (1994)	10	907.0	1.08
Franchi and Montelaghi (1996)	40	898.4	0.12
Saje et al. (1997)	8 (7 internal nodes)	897.2	0.00
*Present	8 (1 internal node)	933.1	4.00
Present	10 (1 internal node)	906.2	1.00
Present	16 (1 internal node)	897.2	0.00

In contrast, the present study demonstrated that accurate results could be efficiently obtained with as few as 10 elements. For the 16-element discretization, the results align closely with the reference solution by Da Deppo and Schmidt (1975), confirming the effectiveness of this approach.

The following problem involves bending a semi-circular, cantilevered arch, depicted in Fig. 3.20. Under an applied moment $M_0 = 1$, the flexible arch fully straightens. With further increases in the bending moment, the arch reforms into a circular shape at a moment of $M_0 = 3$. At this circular configuration, the slope deviation was within 1% error. The final benchmark examines the postbuckling behavior of a shallow arch, where the VOSM formulation successfully captured the characteristic looping deformations along the postbuckling path Fig. 3.21.

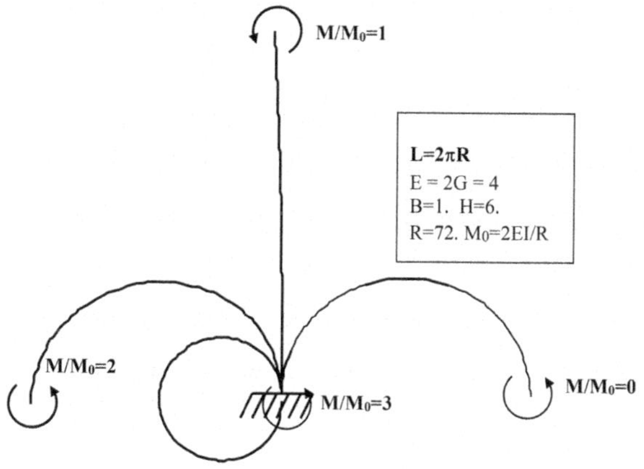

Fig. 3.20 Large deflections of a flexible arch subjected to end moment

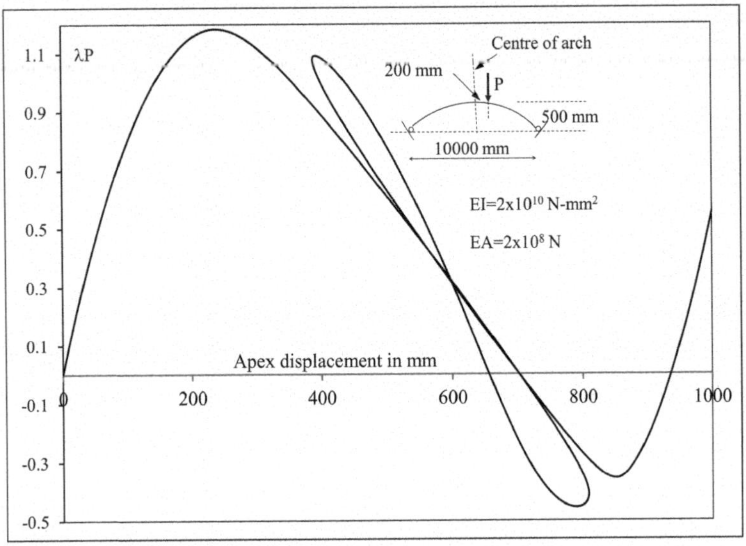

Fig. 3.21 Looping deformations in the postbuckled state of a shallow arch

3.7 Summary

This chapter introduces the Variable Order Secant Matrix (VOSM) method, a novel approach for kinematically exact postbuckling analysis of plane beams and frames. Based on Reissner's (1972) exact strain–displacement relations, the VOSM method accounts for axial, shear, and flexural strains in large deformation problems. Expanding the system's strain energy using a Taylor series and comparing it with the secant matrix symbolism introduced by Mallet and Marcal (1968) provide the basis for deriving variable-order secant matrices.

The VOSM method requires at least 16 secant matrices to accurately capture large displacements and rotations in beams and frames. For cases involving extreme deformation, such as a cantilever beam undergoing full circular bending, over 24 secant matrices were required to maintain accuracy, demonstrating the method's suitability for postbuckling analysis in structures with significant deformation and rotation.

A notable advantage of this formulation is its capacity to model an entire frame member using a single beam element, which balances computational efficiency with accuracy. Including multiple higher-order secant matrices adds minimal computational overhead yet provides highly accurate results. This efficiency makes the VOSM method a practical tool for engineers and researchers handling complex structural problems involving large deformations and postbuckling responses.

Appendix

Explicit variable order secant matrices (VOSM) for the kinematically exact analysis of beams and frames $\left(\overline{\mathbf{U}}_{\mathbf{ij}} = \overline{\mathbf{U}}_{\mathbf{ji}}(\mathbf{3} \times \mathbf{3})\right)$

$$\overline{U}(1,1) = \frac{EA(-1)^n\alpha^{2n}}{(2n!)} + \frac{GA_s(-1)^{n+1}\alpha^{2n}}{2n!}$$

$$\overline{U}(1,2) = \frac{EA(-1)^{n-1}\alpha^{2n-1}}{(2n-1)!}$$

$$\begin{aligned}\overline{U}(1,3) = {} & \frac{\left(EAu_{,x} + GA_s v_{,x}\right)(2n)(-1)^n\alpha^{2n-1}}{(2n!)} \\ & + \frac{EA \times 2(2n)\ (-1)^n\alpha^{2n-1}}{(2n!)} + (EA - GA_s) \\ & \frac{2v_{,x}(-1)^{n-1}\ (2n-1)\alpha^{(2n-2)}}{(2n-1)!} + \frac{EA\ \times 2(-1)^{n+1}(2n)\theta^{(2n-1)}}{2n!} \\ & + GA_s\frac{2(-1)^{n-1}(2n)\alpha^{(2n-1)}}{(2n!)}\end{aligned}$$

$$\overline{U}(2,2) = \frac{EA(-1)^{n+1}\alpha^{2n}}{(2n!)} + \frac{EA(-1)^{n-1}\alpha^{2n-1}}{(2n-1)!}$$

$$\begin{aligned}\overline{U}(2,3) = {} & EA\frac{2v_{,x}(-1)^{n+1}(2n)\alpha^{(2n-1)}}{(2n!)} \\ & + \frac{(EA - GA_s)2(-1)^{n-1}(2n-1)\alpha^{2n-2}}{(2n-1)!} \\ & + \frac{(EA - GA_s) \times 2u_{,x}(-1)^{n-1}(2n-1)\alpha^{(2n-2)}}{(2n-1)!} \\ & + \frac{EA \times 2(-1)^n\ (2n-1)\ \theta^{2n-2}}{(2n-1)!}\end{aligned}$$

$$\begin{aligned}\overline{U}(3,3) = {} & \frac{2(-1)^n 2n(2n-1)\alpha^{(2n-2)}}{(2n!)} \\ & + \frac{\left(EAu_{,x}^2 + GA_s v_{,x}^2\right)2(2n)(2n-1)(-1)^n\alpha^{2n-2}}{(2n!)} \\ & + \frac{EA \times 4u_{,x}(-1)^n(2n)(2n-1)\alpha^{2n-2}}{(2n!)} \\ & + \frac{2\left(EAu_{,x}^2 + GA_s u_{,x}^2\right)(-1)^{n+1}(2n)(2n-1)\alpha^{2n-2}}{(2n!)}\end{aligned}$$

$$+\frac{4(EA - GA_s)u_{,x}(-1)^{n-1}(2n-1)(2n-2)\alpha^{2n-3}}{(2n-1)!}$$
$$+\frac{4(EA - GA_s)u_{,x}v_{,x}(-1)(2n-1)(2n-2)\alpha^{2n-3}}{(2n-1)!}$$
$$+\frac{EA(2u_{,x}(-1)^{n}(2n-1)(2n-2)\theta^{2n-3}}{(2n-1)!}$$
$$+\frac{EA \times 2(-1)^{n-1}(2n)(2n-1)\theta^{2n-2}}{(2n!)}$$
$$+\frac{EA \times 2u_{,x}(-1)^{n+1}(2n)(2n-1)\theta^{2n-2}}{2n!}$$
$$+\frac{GA_s 2(-1)^{n+1}(2n)(2n+1)\theta^{(2n-2)}}{(2n!}$$
$$+\frac{4GA_s u_{,x}(-1)^{n+1}(2n)(2n-1)\alpha^{2n-2}}{2n!}$$

The other terms are symmetric.

References

Antman SS (1968) General solution of plane extensible elastica having nonlinear stress–strain laws. Q Appl Math 26:33–47. https://doi.org/10.1090/QAM/99868

Antman SS (1972) Kirchoff's problem for nonlinear elastic rods. Q Appl Math 32:221–240. https://doi.org/10.1090/qam/667026

Argyris JH (1982) Finite element analysis of two and three dimensional elasto plastic frames—the natural approach. Comput Meth Appl Eng 35:221–248. https://doi.org/10.1016/0045-7825(82)90135-9

Belytschko T, Hseih BJ (1973) Nonlinear transient finite element analysis with convected coordinates. Int J Num Meth Eng 7:255–271. https://doi.org/10.1002/nme.1620070304

Bergan PG, Soreide TH (1973) A comparative study of different numerical solution techniques as applied to nonlinear structural problems. Comp Meth Appl Mech Eng 2:185–201. https://doi.org/10.1016/0045-7825(73)90014-5

Boswell LF (1985) A small strain large rotation theory and finite element formulation of thin curved lattice members. Proceedings of Third International Conference on Space Structures. H.Nooshin (Ed), Elsevier, 375–380

Chajes A , Churchill JE (1987) Nonlinear frame analysis by finite element methods. J Struct Engg ASCE 113(6):1221–1235

Chakraborty S, Mazumdar S (1997) An arbitrary shaped arch element for large deflection analysis. Comput Struct 65(4):593–600. https://doi.org/10.1016/S0045-7949(97)80863-2

Chan SL, Zhou ZH (2000) Nonlinear integrated design and analysis of skeletal structures by one element per member. Eng Struct 22:246–257. https://doi.org/10.1016/S0141-0296(98)00090-X

Crisfield MA (1990) A consistent co-rotational formulation for nonlinear three dimensional beam element. Comp Meth Appl Mech Eng 81:131–150. https://doi.org/10.1016/0045-7825(90)90106-V

Deppo DAD, Schmidt LA (1975) Instability of clamped—hinged circular arches subjected to a point load. J Appl Mech ASME 97:894–896. https://doi.org/10.1115/1.3423734

Duxbury PG, Crisfield MA, Hunt GW (1989) Benchmark for geometric nonlinearity. Comput Struct 33:21–29. https://doi.org/10.1016/0045-7949(89)90125-9 Get rights and content

Eiki Y, Nukulchai WK, Hammadeh M, Kubo Y (1999) Simple degenerate formulation for large displacement analysis of beams. J Eng Mech ASCE 125:1140–1146. https://doi.org/10.1061/(ASCE)0733-9399(1999)125:10(1140)

Epstein M, Murray DW (1976) Large deformation in-plane analysis of elastic beams. Comput Struct 6, 1–9

Franchi CG, Montelaghi F (1996) A weak-weak formulation for large displacement beam statics. A finite volume approximation. Int J Num Meth Eng 39:585–604. https://doi.org/10.1002/(SICI)1097-0207(19960229)39:4%3c585::AID-NME871%3e3.0.CO;2-F

Frisch-Fay R (1961) A new approach to the analysis of the deflection of thin cantilevers. J Appl Mech ASME 28:87–90

Goto Y, Tomow YM, Makoto O (1990) Elliptic integral solution of plane elastica with axial and shear deformation. Int J Solids Struct 20(4):375–390

Gummadi LNB, Palazotto AN (1998) Large strain analysis of beams and arches undergoing large rotations. Int J Nonlinear Mech 33:615–645. https://doi.org/10.1016/S0020-7462(97)00033-4

Hsiao KM, Hou FY (1987) Nonlinear finite element analysis of elastic frames. Comput Struct 26:693–701. https://doi.org/10.1016/0045-7949(87)90016-2

Huddleston JV (1968) Finite deflection and snap-through of high circular arches. J Appl Mech ASME 4, 763–769

Huddleston JV (1967) Finite deflection and snap-through of a two member frame. Int J Solids Struct 3:1023–1030. https://doi.org/10.1016/0020-7683(67)90027-3

Ibrahimbegovic A (1995) On finite element implementation of geometrically nonlinear Reissner's beam theory: three dimensional curved beam elements. Com Meth Appl Mech Eng 122:11–26. https://doi.org/10.1016/0045-7825(95)00724-F

Ibrahimbegovic A, Frey F (1993) Finite element analysis of linear and nonlinear planer deformation of elastic initially curved beams. Int J Num Meth, Eng 36:3239–3258. https://doi.org/10.1002/nme.1620361903

Jennings A (1968) Frame analysis including change of geometry. J Struct Eng ASCE 94:627–644. https://doi.org/10.1061/JSDEAG.0001908

Khamlichi A, Elbakkali L, Limam A (1998) Postbuckling of elastic beams considering higher order strain terms. J Eng Mech ASCE 127:372–378. https://doi.org/10.1061/(ASCE)0733-9399(2001)127:4(372)

Li M (1997) The finite deformation theory for beam, plate and shell Part I. The two-dimensional beam theory. Comp Methods Appl Mech Eng 146:53–63. https://doi.org/10.1016/S0045-7825(96)01213-3

Mallet RH, Marcal PV (1968) Finite element analysis of nonlinear structures. J Struct Div ASCE 94:2081–2105. https://doi.org/10.1061/JSDEAG.0002066

Mattiasson K (1981) Numerical results from large deflection beam and frame problem analysed by means of elliptic integrals. Int J Num Meth Eng 17:145–153. https://doi.org/10.1002/nme.1620170113

Mattiasson K, Bengtsson A, Samuelsson A (1986) On the accuracy and efficiency of numerical algorithm for geometrically nonlinear structural analysis. In. Finite element methods for nonlinear problems, Europe-US symposium, Trandheium, Norway, Bergan, Bathi, Wunderlich, pp 1–23. https://doi.org/10.1007/978-3-642-82704-4_1

Mattiasson K, Samuelsson A (1984) Total and updated Lagrangian forms of the co-rotational finite element formulations in geometric and material nonlinear analysis. In: Taylor et al (ed) Proceeding of international conference on numerical methods for nonlinear problems. Pineridge Press, Swansea

Morteza AM, Torkamani SM, Cao J (1997) Second order elastic plane frame analysis using finite element method. J Struct ASCE 123:1225–1235. https://doi.org/10.1061/(ASCE)0733-9445(1997)123:9(1225)

Oran C, Kassimali A (1976) Large deformation of framed structures under static and dynamic loads. Comput Struct 6:539–547. https://doi.org/10.1016/0045-7949(76)90050-X
Pacoste C, Eriksson A (1995) Element behaviour in post-critical frame analysis. Comp Meth Appl Mech Eng 125:319–343
Pajunen S (1999) Fully nonlinear analysis of plane frame using kinematically exact beam elements. In. Mikkola M, Kouhia R NSCM-12. Proceedings M, pp 163–166
Pajunen S (2000) Large deflection elastoplastic analysis of beams using kinematically exact element. Commun Num Meth Eng 16:407–504. https://doi.org/10.1002/1099-0887(200007)16:7%3c497::AID-CNM352%3e3.0.CO;2-B
Petrolito J, Legge KA (1995) Benchmarks for nonlinear elastic frame analysis. Steel Construction AISC 29(1), 2–12
Petrolito J, Legge KA (1996) Unified nonlinear elastic frame analysis. Comput Struct 60:21–30. https://doi.org/10.1016/0045-7949(95)00378-9
Pi YL, Trahair NS (1996) Buckling strength of steel arches. Proceedings of third Asian-Pacific conference on computational mechanics. Seoul, Korea. https://doi.org/10.1061/(ASCE)0733-9445(1996)122:7(734)
Pi YL, Trahair NS (1998) Nonlinear buckling and postbuckling of elastic arches. Eng Struct 20:571–579. https://doi.org/10.1016/S0141-0296(97)00067-9
Pignataro MA Di Carlo, Casuaro R (1982) On nonlinear beam model from the point of view of computational postbuckling analysis. Int J Solids Struct 18(4), 321–347
Pignatro M, Rizzi N, Luango A (1991) Stability of bifurcation and post critical behaviour of elastic structures. Elsevier Publishers, NY
Rajasekaran S, Murray DW (1973) Incremental finite element matrices. J Struct Eng ASCE 99:2423–2438. https://doi.org/10.1061/JSDEAG.0003667
Ramesh G, Krishnamoorthy CS (1993) Postbuckling anlaysis of structures by dynamic relaxation. Int J Num Meth Eng 36:1339–1364. https://doi.org/10.1002/nme.1620360806
Reddy JN (1997) On locking free shear deformable beam finite element. Comp Meth Appl Mech Eng 149:113–132. https://doi.org/10.1016/S0045-7825(97)00075-3
Reissner E (1972) On one dimensional finite-strain beam theory, the plane problem. J App Math Phys 23:795–804. https://doi.org/10.1007/BF01602645
Salerno G, Lanzo AD (1997) A nonlinear beam finite element for the postbuckling analysis of plane frames by Koiter's perturbation approach. Comp Meth Appl Mech Engg 146, 325–349
Saje M (1990) A variational principle for finite planar deformation of straight slender elastic beams. Int J Solids Struct 26:887–900. https://doi.org/10.1016/0020-7683(90)90075-7
Saje M (1991) Finite element formulation of finite planar deformation of curved elastic beams. Comput Struct 39:327–332. https://doi.org/10.1016/0045-7949(91)90030-P
Saje M, Planni I, Turk G, Vratanar B (1997) A kinematically exact finite element formulation of planar elastic-plastic frames. Comp Meth Appl Mech Eng 144:125–151. https://doi.org/10.1016/S0045-7825(96)01172-3
Shakourzadeh H, Guo YG, Batoz JL (1999) Modelling of connections in the analyses of thin walled space frames. Comput Struct 71:423–433. https://doi.org/10.1016/S0045-7949(98)00241-7
Teh LH (2001) Cubic beam elements in practical analysis and design of steel frames. Eng Struct 23:1243–1255. https://doi.org/10.1016/S0141-0296(01)00032-3
Thompson JMT, Hunt GW (1973) A general theory of elastic stability. Wiley. https://doi.org/10.1016/0022-5096(63)90003-6
Wood RD, Zienkiewicz OC (1977) Geometrically nonlinear finite element analysis of beams, frames arches and axisymmetric shells. Comput Struct 7, 725–715
Yang YB, Chiou HT (1986) Rigid body motion test for nonlinear analysis with beam elements. J Eng Mech ASCE 9:1404–1419. https://doi.org/10.1061/(ASCE)0733-9399(1987)113:9(1404)

Chapter 4
Variable Order Secant Matrices for Geometric Nonlinear Analysis of Thin-Walled Open Sections

Abstract This chapter introduces a computational framework utilizing Variable Order Secant Matrices (VOSM) for the nonlinear geometric analysis of thin-walled open sections. Thin-walled members, known for their high strength-to-weight ratios, are crucial in structural engineering but present challenges due to their complex flexural–torsional coupling behavior under large deformations. Using shell theory, a 3D beam finite element with 14 degrees of freedom is formulated to capture axial, flexural, shear, and torsional deformations. The strain energy of these elements is expanded through the Taylor series to develop higher-order secant stiffness matrices, enhancing accuracy in the large deformation analysis. The VOSM approach is validated through classical examples of pre and postbuckling behaviors under axial loads and end moments, demonstrating both accuracy and computational efficiency. This method proves highly effective for modeling the nonlinear behavior of slender, thin-walled sections subjected to combined torsional and flexural loads.

Keywords Variable order secant matrices · Thin-walled open sections · Geometric nonlinear analysis · Torsional-flexural coupling · Finite element method · Shell theory · Postbuckling analysis

4.1 Earlier Research Work on Thin-Walled Open Sections

Thin-walled open sections, with their high strength-to-weight ratios and efficient use of materials, are pivotal in structural engineering. There has been continued research on the use of thin-walled open sections in structural practice. The research on the analysis of thin-walled open sections is briefly reviewed here.

The shell theory has been widely used to study the critical loads and nonlinear behavior of thin-walled members. Horne et al. (1971) analyzed doubly symmetric open sections, deriving analytical expressions and comparing them with numerical solutions. Lee and Harris (1979) introduced formulations using the incremental

A. Jayachandran, *Elastic Postbuckling Analysis of Structural Systems*, Solid Mechanics and Its Applications 232, https://doi.org/10.1007/978-3-031-89936-2_4

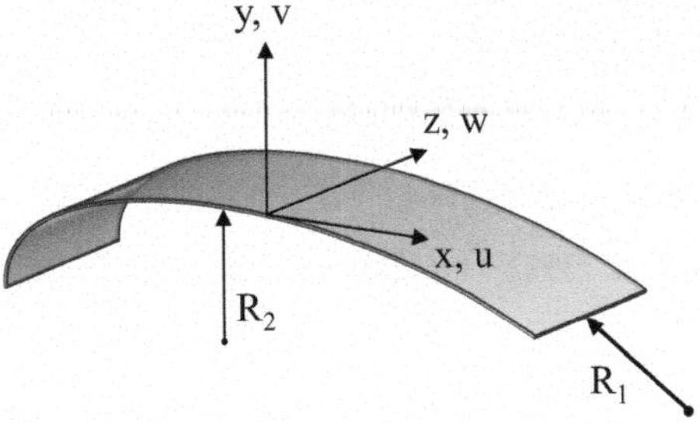

Fig. 4.1 Thin-shell member

variational principle with an efficient tangent stiffness matrix for postbuckling analysis. Trahair and Bild (1990) addressed nonlinear biaxial bending and torsion of thin-walled members, deriving equilibrium equations from energy principles.

Attard (1986) developed a nonlinear theory of non-uniform torsion based on finite displacements, while Wekezer (1985) used membrane shell theory with internal constraints for thin-walled sections, combining analytical and experimental results through the finite element method. Laudiero and Zaccaria (1988) formulated a shell model for thin-walled open sections that incorporated torsional-extensional coupling and geometric effects of applied torque.

Roberts and Azizian (1983) provided a closed-form solution for the lateral buckling of beams subjected to uniform bending. Shakourzadeh et al. (1995) introduced a thin-walled beam finite element based on Benscoter's beam theory. Gendy et al. (1992) addressed shear/warping locking using a mixed variational principle and presented numerical simulations to validate their approach. Gotluru et al. (2000) presented geometric nonlinear analysis methods validated through finite element and experimental comparisons.

Mohri et al. (2002) explored nonlinear postbuckling analysis of thin-walled open-section beams, considering bending and torsion coupling. Si-Wei Liu et al. (2019) and Gao et al. (2021) proposed improved line-element formulations for simulating axial buckling and dynamic responses of thin-walled steel members. Abdelrahman et al. (2022) further advanced non-prismatic line-element formulations, incorporating shear deformations and warping effects.

Barsoum and Gallagher (1970) contributed to the derivation of stiffness matrices for torsional and lateral stability in flexural members. Rajasekaran (1977; 1994) and Rajasekaran and Murray (1974) provided significant advancements in the analysis of tapered thin-walled beams, using the Secant matrix method for large deflection analysis.

In this chapter, the nonlinear theory of shells is used to derive a 3D beam element, capturing higher-order axial, flexural, shear, and torsional deformations of a thin-walled beam. The use of series expansion of strain energy leads to the derivation of variable order secant matrices (VOSM), which are used to formulate a Total Lagrangian (TL) -based 14 degrees-of-freedom 3D beam finite element. The details are presented here.

4.2 Theory of Thin Shells

Consider the geometry of a thin shell, as shown in Fig. 4.2 x, y, z represent a curvilinear frame of reference, with x and z representing the directions of principal curvatures of the mid-surface of the shell in its unloaded state. y represents the direction normal to the mid-surface. R_1 and R_2 are the principal radii of curvature of the shell. The displacements u, v, and w in x, y, and z directions, respectively, are related to the Green's strain tensor $[\varepsilon]$ as:

$$[\varepsilon] = \begin{bmatrix} \varepsilon_{xx} & \varepsilon_{xy} & \varepsilon_{xz} \\ \varepsilon_{yx} & \varepsilon_{yy} & \varepsilon_{yz} \\ \varepsilon_{zx} & \varepsilon_{zy} & \varepsilon_{zz} \end{bmatrix}, \tag{4.1}$$

where

$$\varepsilon_{xx} = \frac{\partial u}{\partial x} + \frac{v}{R_2} + \frac{1}{2}\left[\left(\frac{\partial w}{\partial x}\right)^2 + \left(\frac{\partial u}{\partial x} + \frac{v}{R_2}\right)^2 + \left(\frac{\partial v}{\partial x} - \frac{u}{R_2}\right)^2\right]$$

$$\varepsilon_{zz} = \frac{\partial w}{\partial z} + \frac{v}{R_1} + \frac{1}{2}\left[\left(\frac{\partial w}{\partial z} + \frac{v}{R_1}\right)^2 + \left(\frac{\partial u}{\partial z}\right)^2 + \left(\frac{\partial v}{\partial z} - \frac{w}{R_1}\right)^2\right].$$

$$\varepsilon_{xz} = \frac{1}{2}\left(\frac{\partial u}{\partial z} + \frac{\partial w}{\partial x}\right) + \frac{1}{2}\left[\frac{\partial w}{\partial x}\left(\frac{\partial w}{\partial z} + \frac{v}{R_1}\right) + \frac{\partial u}{\partial z}\left(\frac{\partial u}{\partial x} + \frac{v}{R_2}\right) + \left(\frac{\partial v}{\partial z} - \frac{w}{R_1}\right)\left(\frac{\partial v}{\partial x} - \frac{u}{R_2}\right)\right]$$

The strain energy per unit area of a thin shell is given by:

$$U = \frac{Et}{2(1-\upsilon^2)}\left[(\varepsilon_{xx} + \varepsilon_{zz})^2 - 2(1-\upsilon)\left(\varepsilon_{xx}\varepsilon_{zz} - \frac{1}{4}\varepsilon_{xz}^2\right)\right] + \frac{Et^3}{24(1-\upsilon^2)}\left[(\chi_{xx} + \chi_{zz})^2 - 2(1-\upsilon)\left(\chi_{xx}\chi_{zz} - \tau^2\right)\right], \tag{4.2}$$

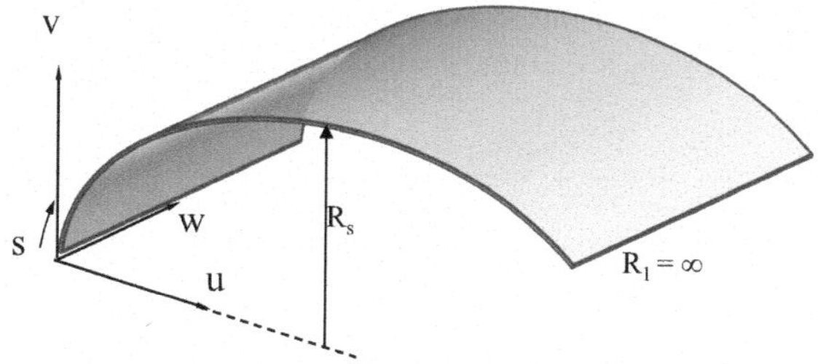

Fig. 4.2 Schematic diagram of a thin-walled open-section member

where E, v, and t are Young's modulus, Poisson's ratio, and the shell thickness, respectively. χ_{xx} and χ_{zz} are the curvatures along x and z, respectively, and τ is the cross-section twist of the member. To apply the theory of thin shells to a thin-walled beam element, we make the following assumptions: The beam is initially straight, and the beam cross-section has a finite radius of curvature ($R_1 = \infty$ and $R_2 = R_s$).

The ratios $\frac{t}{d} \leq 0.1$, $\frac{d}{L} \leq 0.1$ are to be satisfied to classify the beam member as a thin shell, where L is the member length and d is the characteristic width of the cross-section.

The cross-section of the shell is rigid, i.e., $\varepsilon_{ss} = 0$, and $\chi_{ss} = 0$, where s is the coordinate taken along the cross-section of the thin-walled member. The shearing deformations of the middle surface are assumed to be zero, i.e., $\varepsilon_{zs} = 0$ and χ_{ss} can be neglected as long as there are no direct concentrated loads on the shell surface. The strain energy can be written as

$$U = \frac{Et}{2(1-\upsilon^2)}\left[\varepsilon_{zz}^2\right] + \frac{Et^3}{12(1+\upsilon)}\left[\tau^2\right]. \tag{4.3}$$

Thus, the cross-section twist τ is given by:

$$\tau = \frac{-\partial^2 v}{\partial x \partial z} + \frac{\partial}{\partial z}\left(\frac{u}{R_s}\right). \tag{4.4}$$

4.3 Thin-Walled Open-Section Beam Kinematics

The transverse displacements u, v at any point on the cross-section of the thin-walled member are determined using the kinematics of rigid body as (Fig. 4.3):

$$\begin{aligned} u &= u_s - \left(y - e_y\right)\sin\phi - (x - e_x)(1 - \cos\phi), \\ v &= v_s + (x - e_x)\sin\phi - \left(y - e_y\right)(1 - \cos\phi), \end{aligned} \tag{4.5}$$

where u_s and $v_{s.}$ are displacements of the shear center, e_x and e_y are the coordinates of the shear center measured from the origin, and ϕ is the angle of twist. Out-of-plane displacement w cannot be directly expressed in terms of u_s, v_s, and ϕ.

$$\begin{aligned} &\frac{\partial u}{\partial z} + \frac{\partial w}{\partial s} + \frac{\partial w}{\partial s}\left(\frac{\partial w}{\partial z} + \frac{v}{R_1}\right) + \frac{\partial u}{\partial z}\left(\frac{\partial u}{\partial s} + \frac{v}{R_2}\right) \\ &\quad + \left(\frac{\partial v}{\partial z} - \frac{w}{R_1}\right)\left(\frac{\partial v}{\partial s} - \frac{u}{R_2}\right) = 0. \end{aligned}$$

$$\frac{1}{R_1} = 0;\quad \frac{\partial w}{\partial z} = 0;\quad \frac{\partial u}{\partial z} + \frac{\partial w}{\partial s} + \frac{\partial u}{\partial z}\left(\frac{\partial u}{\partial s} + \frac{v}{R_s}\right) + \frac{\partial v}{\partial z}\left(\frac{\partial v}{\partial s} - \frac{u}{R_s}\right) = 0$$

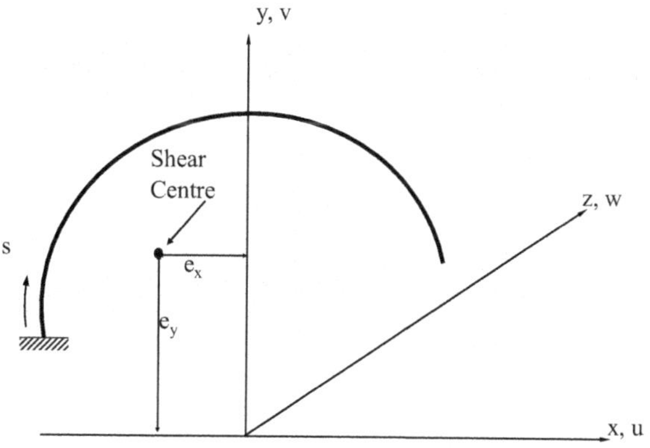

Fig. 4.3 Cross-section kinematics of a thin-walled member

$$\frac{\partial w}{\partial s} = -\frac{\partial u}{\partial z}\left(1+\frac{\partial u}{\partial s}+\frac{v}{R_s}\right)-\frac{\partial v}{\partial z}\left(\frac{\partial v}{\partial s}-\frac{u}{R_s}\right)$$
$$w(z,s) = w_c(z) + \int_0^s \left[-\frac{\partial u}{\partial z}\left(1+\frac{\partial u}{\partial s}+\frac{v}{R_s}\right)-\frac{\partial v}{\partial z}\left(\frac{\partial v}{\partial s}-\frac{u}{R_s}\right)\right].ds \tag{4.6}$$

w_c is the axial displacement of the centroid. To express u and v in terms of s, we define another coordinate system, t, and h, in the cross-section. $t(s)$ is the tangent at any point on the cross-section, and $h(s)$ is the normal from the shear center to the tangent $t(s)$. $\alpha(s)$ is the angle $t(s)$ makes with the x-axis.

$$h(s) = (x-e_x)\sin(\alpha) + \left(y-e_y\right)\cos(\alpha),$$
$$t(s) = (x-e_x)\cos(\alpha) - \left(y-e_y\right)\sin(\alpha), \tag{4.7}$$

To perform the integration in Eq. (4.6), we note that

$$t(s)\cos(\alpha) + h(s)\sin(\alpha) = (x-e_x),$$
$$h(s)\cos(\alpha) - t(s)\sin(\alpha) = \left(y-e_y\right), \tag{4.8}$$

Hence,

$$\cos(\alpha)dt + \sin(\alpha)dh = dx.$$
$$\cos(\alpha)dh - \sin(\alpha)dt = dy. \tag{4.9}$$

But,

$$\cos(\alpha)ds = dx \quad ds = R_s d\alpha \quad \sin(\alpha)ds = -dy \tag{4.10}$$

ds is the profile's arc element, and dx and dy are infinitesimal projections of ds on x and y. We further define

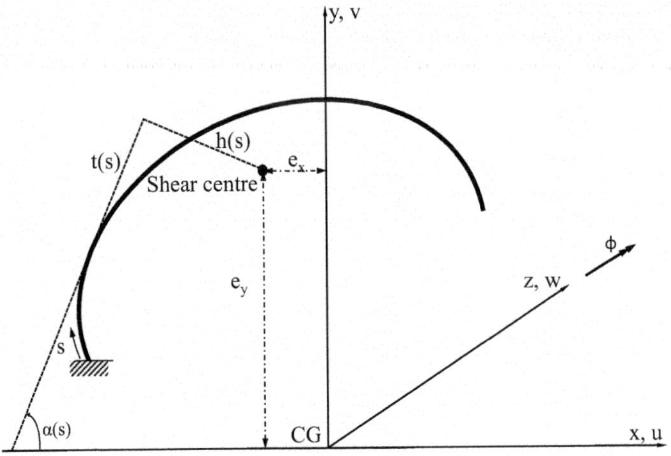

Fig. 4.4 Tangent and normal of the cross-section

$$h\,ds = d\omega \tag{4.11}$$

where $d\omega$ is twice the area of the sector whose base is arc element ds. Then, $\omega(s) = \int_0^s d\omega$ is twice the area of Fig. 4.4 the sector enclosed from 0 to s. This quantity is called the sectorial area. Substituting for u and v in Eq. (4.6),

$$\begin{aligned} w(z,s) = w_c(z) &- \left[v_s'(z)\cos(\phi(z)) + u_s'(z)\sin(\phi(z))\right]y \\ &+ \left[u_s'(z)\cos(\phi(z)) - v_s'(z)\sin(\phi(z))\right]x. \end{aligned} \tag{4.12}$$

The strain energy is,

$$U = \frac{Et}{1-\nu^2}\left(\varepsilon_{zz}^2\right) + \frac{E}{2(1+\nu)}\frac{t^3}{3}\left(\tau^2\right), \tag{4.13}$$

$$\begin{aligned} U = \frac{1}{2}Et \int_l \int_s &\begin{pmatrix} w_c' - \left[v_s''\cos(\phi) + u_s''\sin(\phi)\right]y \\ +\left[u_s''\cos(\phi) - v_s''\sin(\phi)\right]x \\ -\omega\phi'' + \frac{1}{2}\left(u_s'^2 + v_s'^2\right) + \frac{1}{2}\left(h^2 + t^2\right)\phi'^2 \end{pmatrix}^2 ds.dz \\ + G\frac{t^3}{3}\int_l \int_s &\left(\phi'\cos(\phi)\right)^2 ds.dz, \end{aligned} \tag{4.14}$$

where l denotes the total length of the member and $\frac{E}{1-\nu^2} \approx E$, $\frac{E}{2(1+\nu)} = G$. If the x and y axes coincide with the principal axes through the centroid, this imposes orthogonality conditions on the integrals given below:

$$\begin{aligned}
&\int_s 1 \cdot y \cdot ds = 0, \ \int_s 1 \cdot x \cdot ds = 0, \ \int_s 1 \cdot \omega \cdot ds = 0, \\
&\int_s x \cdot y \cdot ds = 0, \ \int_s x \cdot \omega \cdot ds = 0, \ \int_s y \cdot \omega \cdot ds = 0,
\end{aligned} \tag{4.15}$$

The section constants are defined as follows:

$$\begin{aligned}
&t\int_s ds = A, \quad t\int_s x^2 \cdot ds = I_y, \quad t\int_s y^2 \cdot ds = I_x, \\
&t\int_s \omega^2 \cdot ds = I_w, \quad t\int_s \left(h(s)^2 + t(s)^2\right) \cdot ds = I_0, \quad \int_s \frac{t^3}{3} \cdot ds = J,
\end{aligned} \tag{4.16}$$

where A is the cross-section area, I_x and I_y are the principal moments of inertia, I_w is the warping constant, I_0 is the polar moment of inertia, and J is the St. Venant's torsion constant.

Following are additional section constants:

$$\begin{aligned}
&t\int_s \left(h(s)^2 + t(s)^2\right)^2 \cdot ds = \Pi_0, \quad t\int_s \left(h(s)^2 + t(s)^2\right) y \cdot ds = \Pi_x, \\
&t\int_s \left(h(s)^2 + t(s)^2\right)^2 x \cdot ds = \Pi_y, \quad t\int_s \left(h(s)^2 + t(s)^2\right)^2 \omega(s) \cdot ds = \Pi_w.
\end{aligned} \tag{4.17}$$

The expression for strain energy can be expressed as a Taylor series expansion by making the following substitutions:

$$\begin{aligned}
\sin(\phi) &= \phi - \frac{\phi^3}{3!} + \frac{\phi^5}{5!} - \frac{\phi^7}{7!} + \ldots = \sum \left(\frac{(-1)^{n+1}}{(2n-1)!} \phi^{(2n-1)} \right) = a_n \phi^{(2n-1)}, \\
\cos(\phi) &= 1 - \frac{\phi^2}{2!} + \frac{\phi^4}{4!} - \frac{\phi^6}{6!} + \ldots = \sum \left(\frac{(-1)^{n+1}}{(2n-2)!} \phi^{(2n)} \right) = b_n \phi^{(2n)}.
\end{aligned} \tag{4.18}$$

4.4 Variable Order Secant Matrices for Thin-Walled Open Sections

Substituting Eqs. (4.15)-(4.18) in Eq. (4.14), the strain energy is obtained in a polynomial form amenable for the development of Mallet and Marcal's secant stiffness matrices as shown below:

$$U = \int_l \int_s \left\{ \begin{array}{l} EA\left(\frac{w_c'^2}{2} + \frac{w_c u_s'^2}{2} + \frac{w_c v_s'^2}{2} + \frac{u_s'^4}{8} + \frac{v_s'^4}{8} + \frac{u_s'^2 v_s'^2}{4}\right) \\ +\mathrm{EI_y}\left(\begin{array}{l} \frac{u_s'^2}{2} \times \left(b_n\phi^{2n}\right)^2 + u_s' v_s' \times \left(b_n\phi^{2n}\right) \\ \times\left(a_n\phi^{2n-1}\right) + \frac{v_s'^2}{2} \times \left(a_n\phi^{2n-1}\right)^2 \end{array}\right) \\ +EI_x\left(\begin{array}{l} \frac{v_s'^2}{2} \times \left(b_n\phi^{2n}\right)^2 - u_s' v_s' \times \left(b_n\phi^{2n}\right) \\ \times\left(a_n\phi^{2n-1}\right) + \frac{u_s'^2}{2} \times \left(a_n\phi^{2n-1}\right)^2 \end{array}\right) \\ +EI_0\left(\frac{\phi'^2 w_c'}{2} + \frac{\phi'^2 u_s'^2}{4} + \frac{\phi'^2 v_s'^2}{4}\right) + EI_w\left(\frac{\phi''^2}{2}\right) \\ +GJ\left(\frac{\phi'^2 \times \left(b_n\phi^{2n}\right)^2}{2}\right) \\ +E\Pi_0\left(\frac{\phi'^4}{8}\right) + E\Pi_y\left(\begin{array}{l} -\frac{\phi'^2 u_s'' \times \left(b_n\phi^{2n}\right)}{2} \\ -\frac{\phi'^2 v_s'' \times \left(a_n\phi^{2n-1}\right)}{2} \end{array}\right) \\ +E\Pi_x\left(-\frac{\phi'^2 v_s'' \times \left(b_n\phi^{2n}\right)}{2} - \frac{\phi'^2 u_s'' \times \left(a_n\phi^{2n-1}\right)}{2}\right) \\ +E\Pi_w\left(-\frac{\phi'^2 \phi''}{2}\right) \end{array} \right\} .ds.dz \tag{4.19}$$

As the basis for the secant matrix technique, the strain energy at equilibrium is written symbolically as a series,

$$U = U_2 + U_3 + U_4 + \ldots + U_n + \ldots, \tag{4.20}$$

where, as per Mallet and Marcal's formalism, U_2 represents quadratic terms related to elastic stiffness matrix **K**, U_3 represents cubic terms that constitute the first-order displacement matrix **N1**, U_4 represents the quartic terms related to second-order displacement matrix **N2,** and so on. Thus, U_n represents the terms in strain energy expression U with degree n, which finally constitutes the $(n-2)$th order secant matrix $[\mathbf{N_{n\text{-}2}}]$.

The total potential energy can be written as:

$$\prod = U - \mathbf{p}^{\mathrm{T}}\mathbf{q}, \tag{4.21}$$

where **q** consists of generalized nodal forces, and **p** consists of generalized displacements of the structure. Applying the necessary condition of equilibrium, $\frac{\partial \prod}{\partial \mathbf{p}} = 0$, and keeping in mind the order of dependence of terms U_2, U_3, U_4... on **p**, the equilibrium equation can be written as:

$$\mathbf{Kp} + \frac{1}{2}\mathbf{N1p} + \frac{1}{3}\mathbf{N2p} + \frac{1}{4}\mathbf{N3p} + \ldots = \mathbf{q}. \tag{4.22}$$

In the incremental form as discussed in Chap. 1,

$$\mathbf{K\Delta p} + \mathbf{N1\Delta p} + \mathbf{N2\Delta p} + \mathbf{N3\Delta p} + \ldots = \mathbf{\Delta q}. \tag{4.23}$$

In terms of displacement gradients, the strain energy of an element with length l_e can be expressed as

$$U^e = U_2^e + U_3^e + U_4^e + \ldots U_n^e = \int_0^{l_e} \left(\mathbf{g}^{\mathrm{T}} \begin{bmatrix} \frac{0!}{2!}\left[\hat{\mathbf{K}}^{\mathbf{e}}\right] + \frac{1!}{3!}\left[\widehat{\mathbf{N1}}^{\mathbf{e}}\right] + \frac{2!}{4!}\left[\widehat{\mathbf{N2}}^{\mathbf{e}}\right] \\ + \ldots + \frac{(\mathrm{n}-2)!}{\mathrm{n}!}\left[\widehat{\mathbf{N(n-2)}}^{\mathbf{e}}\right] \end{bmatrix} \mathbf{g} \right).\mathrm{dz}, \tag{4.24}$$

where $\mathbf{g}^{\mathrm{T}}$ is the vector of displacement gradients given by:

$$\mathbf{g}^{\mathrm{T}} = \left\{ w_c', u_s', u_s'', v_s', v_s'', \phi, \phi', \phi'' \right\}. \tag{4.25}$$

Hence, element secant stiffness matrices in terms of displacement gradients are given by:

$$\hat{\mathbf{K}}^{\mathbf{e}} = \frac{\partial^2 U_2}{\partial g_i \partial g_j}, \widehat{\mathbf{N1}}^{\mathbf{e}} = \frac{\partial^2 U_3}{\partial g_i \partial g_j}, \widehat{\mathbf{N2}}^{\mathbf{e}} = \frac{\partial^2 U_4}{\partial g_i \partial g_j}, \ldots \tag{4.26}$$

Two typical examples of $\hat{\mathbf{K}}^{\mathbf{e}}$ and $\widehat{\mathbf{N1}}^{\mathbf{e}}$ are given below.

$$\left[\hat{\mathbf{K}}^{\mathbf{e}}\right] = \begin{bmatrix} EA & 0 & 0 & 0 & 0 & 0 & 0 & 0 \\ 0 & 0 & 0 & 0 & 0 & 0 & 0 & 0 \\ 0 & 0 & EI_y & 0 & 0 & 0 & 0 & 0 \\ 0 & 0 & 0 & 0 & 0 & 0 & 0 & 0 \\ 0 & 0 & 0 & 0 & EI_x & 0 & 0 & 0 \\ 0 & 0 & 0 & 0 & 0 & 0 & 0 & 0 \\ 0 & 0 & 0 & 0 & 0 & 0 & GJ & 0 \\ 0 & 0 & 0 & 0 & 0 & 0 & 0 & EI_w \end{bmatrix}, \tag{4.27}$$

$$\left[\hat{\mathbf{N}}_1^{\mathrm{e}}\right] = \begin{bmatrix} 0 & EAu_s' & 0 & EAv_s' & 0 & 0 & EI_0\phi' & 0 \\ EAu_s' & EAw_c' & 0 & 0 & 0 & 0 & 0 & 0 \\ 0 & 0 & 0 & 0 & E(I_y - I_x)\phi & E(I_y - I_x)v_s'' & -E\Pi_y\phi' & 0 \\ EAv_s' & 0 & 0 & EAw_c' & 0 & 0 & 0 & 0 \\ 0 & 0 & E(I_y - I_x)\phi & 0 & 0 & E(I_y - I_x)u_s'' & E\Pi_x\phi' & 0 \\ 0 & 0 & E(I_y - I_x)v_s'' & 0 & E(I_y - I_x)u_s'' & 0 & 0 & 0 \\ EI_0\phi' & 0 & -E\Pi_y\phi' & 0 & E\Pi_x\phi' & 0 & \begin{pmatrix} -E\Pi_w\phi'' - E\Pi_y u_s'' \\ +E\Pi_x v_s'' + EI_0 w_s' \end{pmatrix} & -E\Pi_w\phi' \\ 0 & 0 & 0 & 0 & 0 & 0 & -E\Pi_w\phi' & 0 \end{bmatrix} \tag{4.28}$$

4.5 Implementation of VOSM Method Using Finite Element Method

As shown in Fig. 4.5, a 14-degree-of-freedom 3D beam element is proposed. The linear and cubic Hermite interpolation functions are employed for w_c, u_s, v_s, and ϕ.

$$w_c(z) = \begin{bmatrix} 1 - \frac{z}{l^e} & \frac{z}{l^e} \end{bmatrix} \begin{Bmatrix} w_i \\ w_j \end{Bmatrix}, \tag{4.29}$$

$$u_S(z) = \left[\left(1 - 3\left(\frac{z}{l^e}\right)^2 + 2\left(\frac{z}{l^e}\right)^3\right) \left(\frac{\mathbf{z}}{\mathbf{l^e}} - 2\left(\frac{z}{l^e}\right)^2 + \left(\frac{z}{l^e}\right)^3\right) \left(3\left(\frac{z}{l^e}\right)^2 - 2\left(\frac{z}{l^e}\right)^3\right) \left(\left(\frac{z}{l^e}\right)^3 - \left(\frac{z}{l^e}\right)^2\right) \right] \begin{Bmatrix} u_i \\ \theta_{yi} \\ u_j \\ \theta_{yj} \end{Bmatrix}, \tag{4.30}$$

$$v_S(z) = \left[\left(1 - 3\left(\frac{z}{l^e}\right)^2 + 2\left(\frac{z}{l^e}\right)^3\right) \left(\frac{\mathbf{z}}{\mathbf{l^e}} - 2\left(\frac{z}{l^e}\right)^2 + \left(\frac{z}{l^e}\right)^3\right) \left(3\left(\frac{z}{l^e}\right)^2 - 2\left(\frac{z}{l^e}\right)^3\right) \left(\left(\frac{z}{l^e}\right)^3 - \left(\frac{z}{l^e}\right)^2\right) \right] \begin{Bmatrix} v_i \\ \theta_{xi} \\ v_j \\ \theta_{xj} \end{Bmatrix}, \tag{4.31}$$

$$\phi(z) = \left[\left(1 - 3\left(\frac{z}{l^e}\right)^2 + 2\left(\frac{z}{l^e}\right)^3\right) \left(\frac{\mathbf{z}}{\mathbf{l^e}} - 2\left(\frac{z}{l^e}\right)^2 + \left(\frac{z}{l^e}\right)^3\right) \left(3\left(\frac{z}{l^e}\right)^2 - 2\left(\frac{z}{l^e}\right)^3\right) \left(\left(\frac{z}{l^e}\right)^3 - \left(\frac{z}{l^e}\right)^2\right) \right] \begin{Bmatrix} \phi_i \\ \phi_i' \\ \phi_j \\ \phi_j' \end{Bmatrix}. \tag{4.32}$$

Thus, **g** and **p** are related as:

$$\mathbf{g} = \mathbf{D}\mathbf{p}, \tag{4.33}$$

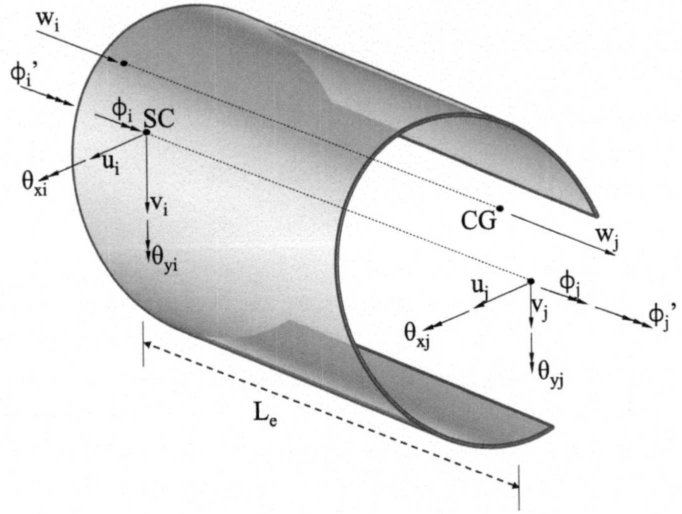

Fig. 4.5 Thin-walled beam finite element with fourteen degrees of freedom

where [**D**] is a matrix that relates the vector of displacement gradients to the nodal degrees of freedom of a 14-degrees-of-freedom beam element.

$$\mathbf{p} = \left\{u_{si}, v_{si}, w_{ci}, \phi_i, \theta_{yi}, \theta_{xi}, \phi_i', u_{sj}, v_{sj}, w_{cj}, \phi_j, \theta_{yj}, \theta_{xj}, \phi_j'\right\}^{\mathrm{T}}. \tag{4.34}$$

Thus, Eq. (4.24) becomes

$$\begin{aligned} \mathrm{U}^{\mathrm{e}} &= \int_0^{\mathrm{l_e}} \left(\mathbf{p}^{\mathrm{T}} \left[\begin{array}{c} \frac{0!}{2!}\mathbf{D}^{\mathrm{T}}\left[\hat{\mathbf{K}}^{\mathrm{e}}\right]D + \frac{1!}{3!}\mathbf{D}^{\mathrm{T}}\left[\widehat{\mathbf{N1}}^{\mathbf{e}}\right]D + \frac{2!}{4!}\mathbf{D}^{\mathrm{T}}\left[\widehat{\mathbf{N2}}^{\mathbf{e}}\right]D \\ +\ldots.. + \frac{(\mathrm{n}-2)!}{\mathrm{n}!}\mathbf{D}^{\mathrm{T}}\left[\widehat{\mathbf{N}(\mathbf{n}-2)}^{\mathbf{e}}\right]D \end{array} \right] \mathbf{p} \right) .\mathrm{dz} \\ &= \mathbf{p}^{\mathrm{T}}\left[\frac{0!}{2!}\mathbf{K}^{\mathrm{e}} + \frac{1!}{3!}\mathbf{N1} + \frac{2!}{4!}\mathbf{N2} + \ldots + \frac{(\mathrm{n}-2)!}{\mathrm{n}!}\mathbf{N}(\mathbf{n}-2)\right]\mathbf{p}, \end{aligned} \tag{4.35}$$

where

$$\begin{aligned} \left[\mathbf{K}^{\mathbf{e}}\right] &= \int_0^{l_e} \left(\mathbf{D}^{\mathrm{T}}\left[\widehat{\mathbf{K}}^{e}\right]\mathbf{D}\right) \cdot dz, \\ \left[\mathbf{N}_{\mathbf{1}}^{e}\right] &= \int_0^{l_e} \left(\mathbf{D}^{\mathrm{T}}\left[\widehat{\mathbf{N1}}^{e}\right]\mathbf{D}\right) \cdot dz, \\ \left[\mathbf{N}_{\mathbf{n-2}}^{e}\right] &= \int_0^{l_e} \left(\mathbf{D}^{\mathrm{T}}[\mathbf{N}(\widehat{\mathbf{n}-2})^{e}]\mathbf{D}\right) \cdot dz. \end{aligned} \tag{4.36}$$

After the stiffness matrix assembly procedure, the total energy of the entire structure is,

$$\prod = U - \mathbf{p}^{\mathrm{T}}\mathbf{q} = \mathbf{p}^{\mathrm{T}}\left[\frac{0!}{2!}\mathbf{K} + \frac{1!}{3!}\mathbf{N1} + \frac{2!}{4!}\mathbf{N1} + \ldots + \frac{(n-2)!}{n!}\mathbf{N}(\mathbf{n}-2)\right]\mathbf{p} - \mathbf{p}^{\mathrm{T}}\mathbf{q}. \tag{4.37}$$

The equilibrium equation can be written as:

$$\left[\mathbf{K} + \frac{1}{2}\mathbf{N1} + \frac{1}{3}\mathbf{N2} + \ldots + \frac{1}{n-1}\mathbf{N}(\mathbf{n}-2)\right]\mathbf{p} = \mathbf{q}. \tag{4.38}$$

K is the linear stiffness matrix, and **N1, N2, N3**, and so on are the secant matrices of higher order. Equation (4.38) is the secant matrix relation used to ascertain the equilibrium during Newton iterations. Considering the displacement dependency of the stiffness matrices, the tangent stiffness relation can be obtained in the incremental form as:

$$[\mathbf{K} + \mathbf{N1} + \mathbf{N2} + \ldots + \mathbf{N(n-2)}]\Delta\mathbf{p} = \Delta\mathbf{q}. \quad (4.39)$$

The Variable Order Secant Matrices appear in the secant and tangent stiffness relations. They can be formed by simply adjusting the scalar multiples. VOSM are series functions that can be expanded to any order.

4.6 Example Problems for Validation

Depending on the order of the stiffness matrices (VoSM) used in the formulation, the secant and tangent stiffness matrices in Eqs. (4.38) and (4.39) capture the effects of large displacements, large torsion, warping, and flexural–torsional coupling. Thus, the finite element model for the 3D thin-walled beam developed in the previous section contains highly coupled and nonlinear secant and tangent equilibrium relations. Hence, iterative methods are necessary for tracing the entire load-deformation response. The capabilities of the proposed beam element in capturing the postbuckling response of thin-walled beams are demonstrated hereafter.

4.6.1 Pre and Postbuckling Behavior of a Simply Supported Beam Column Subjected to Axial Load

This example has been treated by Rajasekaran and Murray (1974). A 100-inch-long simply supported beam of 8WF31 I section subjected to a small constant transverse load (5 kips) at the midspan and a varying uniaxial compressive load, as shown in Fig. 4.6, is investigated. As the system approaches the critical load for this beam, the flexural deformations increase drastically. Using the model with 4 elements, the postbuckling behavior has been performed using a small flexural perturbation at the midspan. The secant stiffness matrices up to order 8 were used in Eq. (4.38) and Eq. (4.39).

The vertical deflection at the mid-node (u) v/s the non-dimensionalized axial load (P/P_{cr}) is shown in Fig. 4.7. The load–displacement path captures the postbuckling response, and the agreement with Rajasekaran and Murray (1974) is good.

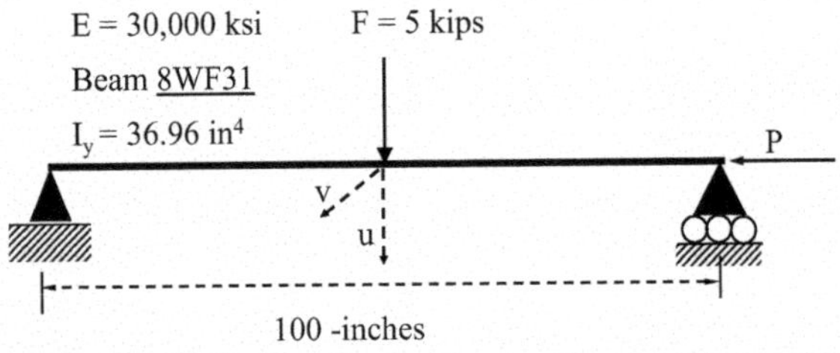

Fig. 4.6 Simply supported beam subjected to axial load and transverse load

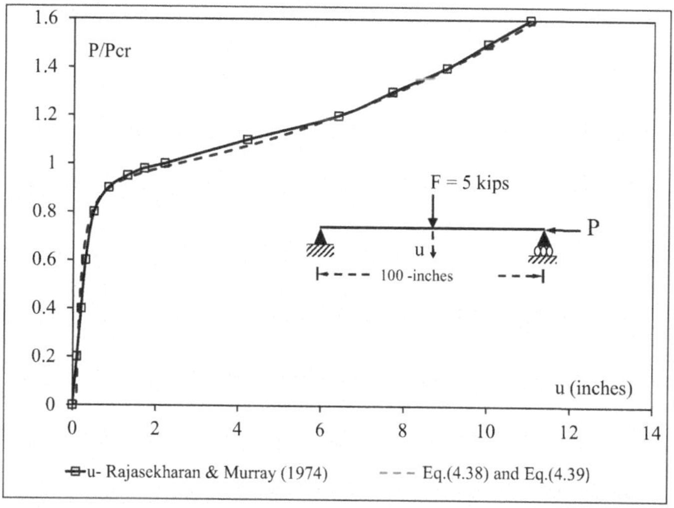

Fig. 4.7 Pre and postbuckling behavior of a simply supported beam column

4.6.2 Simply Supported Beam Column Subjected to End Moments

The last example investigates the lateral stability of a simply supported beam with the *I* section considered with the section dimensions, as shown in Fig. 4.8.

This section's critical lateral torsional buckling moment is computed as 6200 kip-inches. For this cross-section, the postbuckling range exhibits a coupled flexural–torsional behavior, and displacements u, v, and φ are present. The vertical deflection (u), out-of-plane deflection (v), and twist (φ) at the midspan of the beam concerning the applied moment (M) are as given in Figs. 4.9 and 4.10.

Fig. 4.8 Simply supported beam under uniform bending

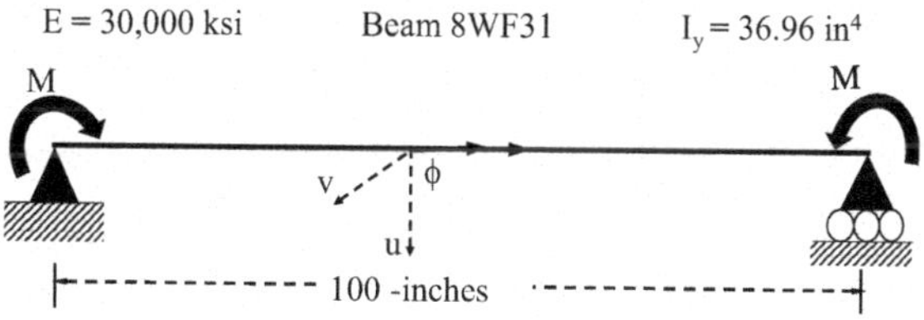

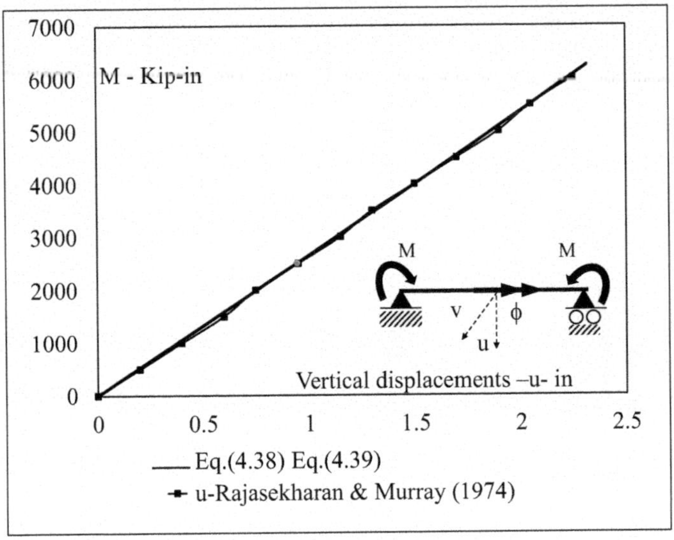

Fig. 4.9 Moment–in-plane vertical displacement (u) graph

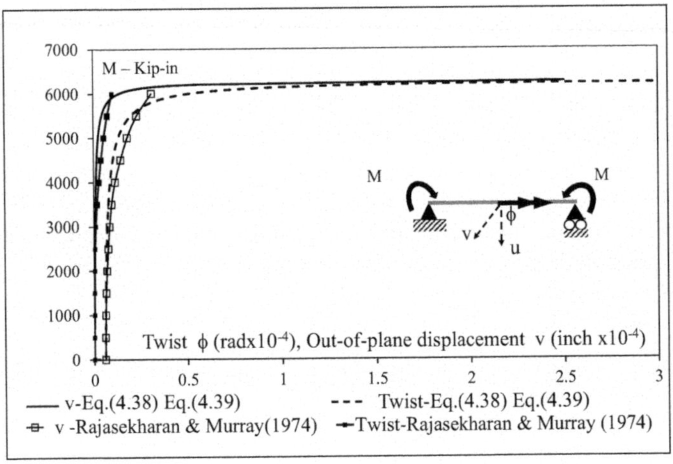

Fig. 4.10 M v/s out-of-plane displacement (v) and twist (φ) graph

4.7 Summary

This chapter proposes a thin-walled beam finite element that can capture large deflection and torsion behavior of thin-walled open sections. The strain energy is formulated using shell theory, and Mallet-Marcal's N1-N2 method is modified and applied to generate the Variable Order Secant Matrices (VOSMs). This computational framework can capture the flexural–torsional coupling in beams using beam elements.

The proposed method is validated using classic examples of pre and post-buckled behavior of beams in flexural buckling and lateral torsional buckling modes.

References

Abdelrahman AHA, Chen L, Liu SW, Ziemian RD (2022) Timoshenko line-element for stability analysis of tapered I-section steel members considering warping effects. Thin-Walled Struct 175:109198. https://doi.org/10.1016/j.tws.2022.109198

Attard MM (1986) Nonlinear theory of non-uniform torsion of thin-walled open beams. Thin-Walled Struct 4–2:101–134. https://doi.org/10.1016/0263-8231(86)90019

Barsoum RS, Gallagher RH (1970) Finite element analysis of torsional and torsional-flexural stability problems. Int J Numer Meth Eng 2–3:335–352. https://doi.org/10.1002/nme.162002 0304

Gao WL, Abdelrahman AHA, Liu SL, Ziemian RD (2021) Second-order dynamic time-history analysis of beam-columns with non-symmetrical thin-walled steel sections. Thin-Walled Struct 160:107367. https://doi.org/10.1016/j.tws.2020.107367

Gendy AS, Saleeb AF, Chang TYP (1992) Generalized thin-walled beam models for flexural-torsional analysis. Comput Struct 42–4:531–550. https://doi.org/10.1016/0045-7949(92)90120

Gotluru BP, Schafer BW, Peköz T (2000) Torsion in thin-walled cold-formed steel beams. Thin-Walled Struct 37–2:27–145. https://doi.org/10.1016/S0263-8231(00)00011-2

Horne MR, Hoh KH, Poskitt TJ (1971) The torsional-flexural behavior of thin-walled prismatic members. Int J Mech Sci 13–07:641–657. https://doi.org/10.1016/0020-7403(71)90034-8

Laudiero L, Zaccaria D (1988) A consistent approach to linear stability of thin-walled beams of open section. Int J Mech Sci 30–7:503–515. https://doi.org/10.1016/0020-7403(88)90004-5

Lee HP, Harris PJ (1979) Post-buckling strength of thin-walled members 10–4:689–702. https://doi.org/10.1016/0045-7949(79)90013-0

Liu SL, Gao WL, Ziemian RD (2019) Improved line-element formulations for the stability analysis of arbitrarily-shaped open-section beam-columns. Thin-Walled Struct 144–1:06290. https://doi.org/10.1016/j.tws.2019.106290

Mohri F, Azrar L, Potier-Ferry M (2002) Lateral post-buckling analysis of thin-walled open section beams. Thin-Walled Struct 40–12:1013–1036. https://doi.org/10.1016/S0263-8231(02)00043-5

Rajasekaran S (1994) Equations for tapered thin-walled beams of generic open section. Eqn Tapered Thin-Walled Beams Generic Open Sect, J Eng Mech ASCE 10–8:1607–1629. https://doi.org/10.1061/(ASCE)0733-9399(1994)120:8(160

Rajasekaran S, Murray DW (1974) Finite element large deflection analysis of thin-walled beams of open sections. In: Pulmono VA, Kabaila AP (eds) International conference on FEM in engineering. New South Wales, Australia, pp 435–454

Rajasekaran S (1977) Finite element method for plastic beam-columns. In: Che WFN, Atsuta T (eds) Theory of beam columns, vol 2. McGraw-Hill, New York, pp 539–608

Roberts TM, Azizian AG (1983) Influence of pre-buckling displacements on the elastic critical loads of thin-walled bars of open cross-section. Int J Mech Sci 25–2:93–104. https://doi.org/10.1016/0020-7403(83)90003-6

Shakourzadeh H, Guo YQ, Batoz JL (1995) A torsion bending element for thin-walled beams with open and closed cross sections. Comput Struct 55–5:1045–1054. https://doi.org/10.1016/0045-7949(94)00509-2

Trahair NS, Bild S (1990) Elastic biaxial bending and torsion of thin-walled members. Thin-Walled Struct 9–1–4:269–307. https://doi.org/10.1016/0263-8231(90)90049-5

Wekezer JW (1985) Nonlinear torsion of thin-walled bars of variable, open cross-sections. Int J Mech Sci 27–10:631–641. https://doi.org/10.1016/0020-7403(85)90045-1

Chapter 5
Improved Secant Matrices for the Postbuckling Analysis of Thin Initially Imperfect Plates

Abstract This chapter presents a secant matrix approach for the postbuckling analysis of thin plates with initial imperfections, which applies to laminated composite and isotropic materials. Using classical laminated plate theory (CLPT), the secant matrices are presented by incorporating the additional displacement formulation (ADF). This proposed secant matrix method with ADF is robust compared to Yang's (1972) secant matrices using total displacement formulation (TDF). The chapter outlines the development of the ADF for constructing relevant secant matrices. The proposed ADF formulation is validated through numerical examples.

Keywords Postbuckling · Secant matrices · Thin imperfect plates · Initial imperfections · Laminated composite plates · Large deflection · Classical laminated plate theory · Additional displacement formulation · Structural stability · Finite element analysis

5.1 Evolution of Secant Matrix Technique for Thin Plates

Thin-walled structures are integral to both aerospace and civil engineering due to their advantageous combination of strength and efficiency. In aerospace applications, thin laminated composite plates are critical, as their material properties can be customized to meet specific strength requirements. In civil engineering, thin plate elements are essential in cold-formed steel structures, offering both lightweight and efficient design solutions. One of the critical design challenges for these thin-walled elements is local buckling under compression, which requires careful analysis and consideration.

The study of local buckling, particularly the linear critical stress at which buckling begins, has been well documented. Leissa (1987) reviewed the buckling behavior of laminated composite plates, highlighting the bifurcation point where disturbances in an axially loaded plate lead to buckling. Unlike slender columns, thin rectangular plates with supported edges continue to deform stably beyond buckling, primarily

A. Jayachandran, *Elastic Postbuckling Analysis of Structural Systems*, Solid Mechanics and Its Applications 232, https://doi.org/10.1007/978-3-031-89936-2_5

due to their postbuckling strength, allowing them to carry additional loads after the critical stress is exceeded.

A critical factor in postbuckling behavior is the presence of imperfections in the plate, which can be comparable to the plate's thickness, as noted by Schafer and Pekoz (1998). These imperfections introduce nonlinearities in the load-deformation of the plate response, illustrating that linear critical stress alone is inadequate for designing thin-walled structures, especially for laminated composites. Although Shen and Williams (1996) and Shen (1997) provided analytical models for the nonlinear behavior of laminated plates, such models become complex when real-world boundary conditions are involved.

The secant matrix method, in particular, is widely employed in nonlinear finite element analysis of postbuckling behavior. Researchers such as Yang (1971), Yang and Han (1983), and Kapania and Yang (1987) applied this method to study plates with initial imperfections. Other studies, including those by Saigal et al. (1986) and Librescu and Souza (1993), have explored the postbuckling of laminated composite plates without relying on the secant matrix approach.

However, explicit secant matrices for postbuckling analysis of initially imperfect laminated or isotropic plates are relatively rare in the literature. This book focuses on developing and validating secant matrices for large deflection and postbuckling analysis of thin laminated composite plates with initial imperfections. It also addresses the shortcomings of previous formulations, particularly their ability to handle the imperfections.

The secant matrix formulations presented in this study are based on classical laminated plate theory (CLPT), which is suitable for many types of laminated composite problems. Additionally, these formulations can be extended to incorporate first-order shear deformation theory (FOST) to account for shear-deformable plates, as demonstrated by Narayanan and Krishnamoorthy (1989) for isotropic plates and Ganapathy and Varadhan (1995) for laminated composites.

5.2 Notable Previous Works on Postbuckling Analysis of Thin Plates

Given their critical roles in structural performance, the postbuckling behavior of thin plates has been explored in both aerospace and civil engineering. Numerous analytical (Rhodes 1986; Shanmugam 1987) and numerical (Sridharan 1982; Barbero and Reddy 1990; Kumar and Kishore 1991; Shin et al. 1993; Sundaresan et al. 1996) studies have significantly advanced our understanding of this phenomenon. For instance, Bradfield (1982) conducted a comprehensive review of the elastoplastic behavior of steel plates, while Chia (1988) and Noor (1995) focused on the buckling behavior of laminated composite plates. During the 1990s, research continued to progress, particularly about large deflection behavior under axial compression (Chang and Sherbourne 1990; Chang and Shabana 1990), highlighting the need for further investigation into the behavior of plates with complex boundary conditions and mixed loading scenarios (Harris and Little 1991; Usami 1993; Paik 1995).

Additionally, interest in the behavior of epoxy-based composite plates subjected to thermo-mechanical loads grew during the 1990s (Librescu et al. 1993; Singh et al. 1994; Sundaresan et al. 1996). Atluri (1997) comprehensively reviewed finite element methods applied to postbuckling in laminated composite panels. However, studies specifically addressing secant matrix formulations for the postbuckling analysis of imperfect laminated plates remain limited. For isotropic plates, Yang (1971) and Yang and Han (1983) made notable contributions, while for laminated composite plates, Ganapathy and Varadhan (1995) and Balamurugan et al. (1996) developed secant matrix techniques, though they did not account for initial imperfections.

The finite strip method has also been applied in postbuckling analysis, with contributions by Dawe et al. (1993) and Lam and Dawe (1998). Other researchers, such as Aravind (1990) and Madasamy (1994), studied large deflection and postbuckling behavior in thin anisotropic plates using classical and spline finite strip methods. Lui and Lam (2001) expanded upon the finite strip method by incorporating polynomial expressions to model initial imperfections, extending its applicability. Though they did not address initial imperfections, Ganapathy and Varadhan (1995) introduced secant matrices for shear-flexible laminated plates.

Previous works by Yang (1971) and Kapania and Yang (1987) did not develop explicit secant matrices for imperfect plates, relying instead on total displacement formulations (TDFs) for postbuckling analysis. However, these were inadequate due to the absence of higher-order displacement approximations. While Kapania and Yang (1987) addressed imperfections through the Love–Kirchhoff thin shell theory, their results for imperfect plates remained inaccurate.

Several other studies have modeled plate imperfections by applying pseudoforces (Mallet and Marcal 1968; Rajasekaran and Murray 1973; Narayanan and Krishnamoorthy 1989; Ganapathy and Varadhan 1995). However, this method does not adequately represent the stress-free nature or the randomness of naturally occurring imperfections. Dow and Smith (1984), Ueda and Tao (1985), and Lui and Lam (2001) argued for the use of a secant matrix approach to more accurately capture random imperfections. The shape of these imperfections significantly impacts the postbuckling response, particularly for laminated composite plates, as noted by Kim et al. (1998) and Lui and Lam (2001).

This review underscores the importance of secant matrix formulations that account for initial imperfections in the postbuckling analysis of isotropic and laminated composite plates as shown in Fig. 5.1.

This study introduces incremental matrices for analyzing the large deflection and postbuckling behavior of thin laminated composite plates with initial curvature. It is demonstrated that the secant matrices for modeling initial imperfections are free from approximations, making them applicable to a wide range of practically observed imperfections in plate structures.

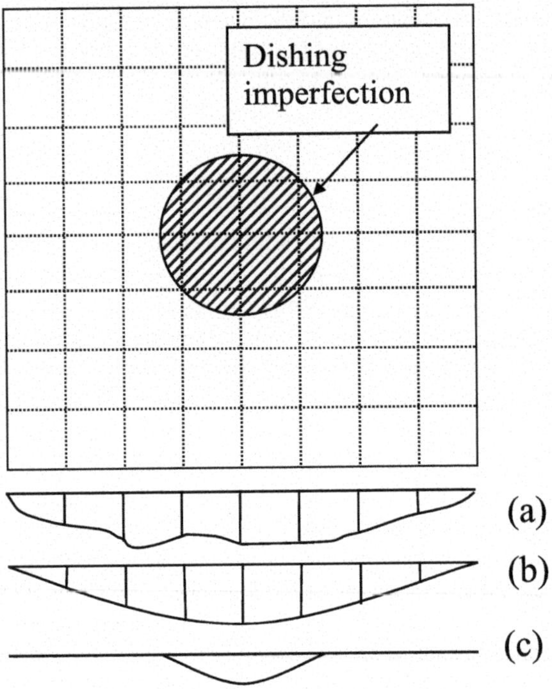

Fig. 5.1 Types of imperfection encountered in postbuckling analysis of plates **a** practically observed **b** idealized **c** localized dishing imperfection

5.3 Improved Secant Matrix Formulations for Initially Imperfect Laminated Composite Plates

The strain energy density U_0 of a Hookian material in a Total Lagrangian (TL) format could be written in terms of the second Piola–Kirchhoff stress $\boldsymbol{\sigma}$ and Green's strain $\boldsymbol{\varepsilon}$ as

$$U_0 = \int \boldsymbol{\sigma} d\boldsymbol{\varepsilon}. \tag{5.1}$$

The deformation of the initially imperfect plate is described in the Total Lagrangian format using a fixed right-handed rectangular Cartesian frame of reference, with the X, Y plane coinciding with the middle surface of the plate in its undeformed state and the Z-axis normal to it as shown in Fig. 5.2. The components of the in-plane and out-of-plane displacements of a particle of the plate initially at (x, y, z) are denoted by $\mathbf{u} = \{u, v, w\}^{\mathrm{T}}$. As the plate is assumed to be very thin and obeys classical laminated plate theory (CLPT) and the traction on surfaces parallel to the middle surface of the plate is negligible, the flexural strains are assumed to vary linearly along the thickness.

Since Marguerre's assumption is valid, the strain terms dependent on the square of the membrane displacements u, v are neglected. Green's strain, based on Marguerre's

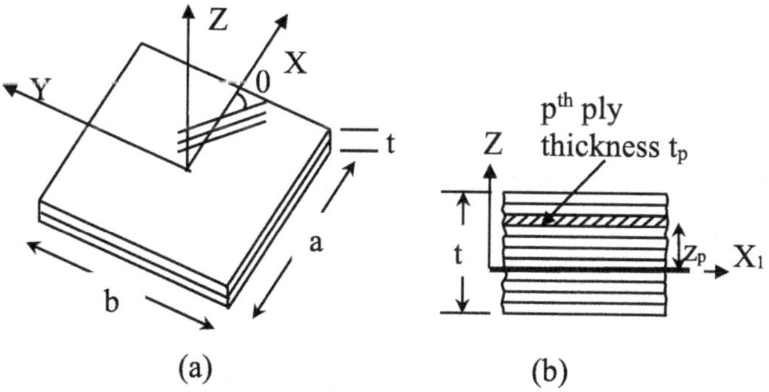

Fig. 5.2 **a** Coordinate axes of the laminated composite plate and ply orientation. **b** Lay-up sequence

shell theory for an initially imperfect plate, could be written in two ways. In the first formulation, called Yang's formulation (Yang 1971) or the total displacement formulation (TDF), Green's strain is written as a function of total displacement, which is the sum of additional displacement 'w' and initial imperfection 'w_0' ($\overline{w} = w + w_0$) as

$$\begin{aligned}
\varepsilon_x &= \frac{\partial u}{\partial x} + \frac{1}{2}\left(\frac{\partial \overline{w}}{\partial x}\right)^2 - \frac{1}{2}\left(\frac{\partial w_0}{\partial x}\right)^2 - z\frac{\partial^2 w}{\partial x^2} \\
\varepsilon_y &= \frac{\partial v}{\partial y} + \frac{1}{2}\left(\frac{\partial \overline{w}}{\partial y}\right)^2 - \frac{1}{2}\left(\frac{\partial w_0}{\partial y}\right)^2 - z\frac{\partial^2 w}{\partial y^2} \\
\gamma_{xy} &= \left(\frac{\partial u}{\partial y} + \frac{\partial v}{\partial x}\right) + \left(\frac{\partial \overline{w}}{\partial x}\frac{\partial \overline{w}}{\partial y}\right) - \left(\frac{\partial w_0}{\partial x}\frac{\partial w_0}{\partial y}\right) - 2z\frac{\partial^2 w}{\partial x \partial y}
\end{aligned} \tag{5.2}$$

In Eq. (5.2), the first term represents the first-order membrane strains. Since the out-of-plane deformations are measured from the undeformed middle plane of the plate, the axial shortening strain is represented in the second term as a quadratic function of the total deflection ($w + w_0$). The last term represents the flexural strains due to the curvature of the plate. By the preceding argument, Eq. (5.2) can be symbolically expressed as

$$\boldsymbol{\varepsilon} = \boldsymbol{\varepsilon}_1^1 + \boldsymbol{\varepsilon}_2^2 - \boldsymbol{\varepsilon}_3^0 - \boldsymbol{\varepsilon}_4^1 \tag{5.3}$$

The superscripts in Eq. (5.3) represent the order of dependence of the strain terms on the displacement gradients. The subscripts represent the first, second, third, and fourth terms in Eq. (5.2). Since the initial imperfection 'w_0' is a constant, it is represented as a zero-order term. This method of representing the strain–displacement relations and subsequent energy formulation is called the total displacement formulation (TDF). Another way of expressing Green's strain for an initially imperfect plate using Marguerre's shell theory is called the additional displacement formulation (ADF). The strain in terms of additional deflection w is given by

$$
\begin{aligned}
\varepsilon_x &= \frac{\partial u}{\partial x} + \frac{1}{2}\left(\frac{\partial w}{\partial x}\right)^2 + \frac{\partial w_0}{\partial x}\frac{\partial w}{\partial x} - z\frac{\partial^2 w}{\partial x^2} \\
\varepsilon_y &= \frac{\partial v}{\partial y} + \frac{1}{2}\left(\frac{\partial w}{\partial y}\right)^2 + \frac{\partial w}{\partial y}\frac{\partial w_0}{\partial y} - z\frac{\partial^2 w}{\partial y^2} \\
\gamma_{xy} &= \left(\frac{\partial u}{\partial y} + \frac{\partial v}{\partial x}\right) + \left(\frac{\partial w}{\partial x}\frac{\partial w}{\partial y}\right) + \left(\frac{\partial w_0}{\partial x}\frac{\partial w}{\partial y} + \frac{\partial w_0}{\partial y}\frac{\partial w}{\partial x}\right) - 2z\frac{\partial^2 w}{\partial x \partial y}.
\end{aligned}
\tag{5.4}
$$

Here Eq. (5.4) can be symbolically written as in Eq. (5.3) as

$$
\boldsymbol{\varepsilon} = \boldsymbol{\varepsilon}_1^1 + \boldsymbol{\varepsilon}_2^2 + \boldsymbol{\varepsilon}_3^1 - \boldsymbol{\varepsilon}_4^1 \tag{5.5}
$$

Here again, the subscripts represent the first, second, third, and fourth terms in Eq. (5.4), and the superscripts the order of displacement gradients. It is shown later that, as far as the secant matrix procedure is concerned, how Green's strain is expressed is extremely important as it decides the range of applicability of the formulation to model initial imperfection. This was pointed out by Arul Jayachandran et al (1995). Since Green's strain $\boldsymbol{\varepsilon}$ is assumed for the description of motion of the plate, its work conjugate, the second Piola–Kirchhoff stress $\boldsymbol{\sigma}$ is used as the stress measure. In the case of isotropic material, the second Piola–Kirchhoff stress can be related to Green's strain using an elastic compliance relation as

$$
\boldsymbol{\sigma} = \mathbf{C}\boldsymbol{\varepsilon}, \tag{5.6}
$$

where $\mathbf{C}$ is a matrix of elastic compliance; alternatively, the displacement gradients in the strain terms in Eq. (5.2) or Eq. (5.4) can be collected into a column vector $\mathbf{g}^{\mathrm{T}}$ and then the second Piola–Kirchhoff stress can be related to the vector of displacement gradient using a modified matrix of material compliance. $\mathbf{C}'$ as,

$$
\boldsymbol{\sigma} = \mathbf{C}'\mathbf{g}. \tag{5.7}
$$

For a layered composite plate (Fig. 5.2b) based on the classical laminated plate theory (CLPT), the material compliance relations can be written in terms of the usual stretching, bending, and coupling rigidities (Datoo 1991) as

$$
\mathbf{C} = \begin{bmatrix} A & B \\ B & D \end{bmatrix} \tag{5.8}
$$

Using the parameters defined in Fig. 5.2, the rigidities can be expressed as (Datoo 1991)

$$
A = \sum_{p=1}^{N} t_p \left(\overline{\mathbf{Q}}\right)_p
$$

$$B = \sum_{p=1}^{N} -t_p \bar{z}_p (\overline{\mathbf{Q}})_p$$

$$D = \sum_{p=1}^{N} \left(t_p \bar{z}_p^2 + \frac{t_p^3}{12} \right) (\overline{\mathbf{Q}})_p \tag{5.9}$$

where N is the number of plys in the lay-up sequence $(\overline{\mathbf{Q}})_p$ is the transformed reduced stiffness term of the pth ply. The strain energy for the domain of the plate Ω in terms of the displacement gradients can be written as

$$U_0 = \int_{\Omega} \varepsilon^{\mathrm{T}} C \varepsilon \, d\Omega \tag{5.10}$$

Applying Eq. (5.2) or Eq. (5.4) and Eq. (5.7) into Eq. (5.10), the elastic strain energy could be obtained as a function of displacement gradients. By examining the order of dependence of Green's strain, the product terms in Eq. (5.10) could be split as strain energy components, which are functions of second, third, and fourth-order of displacement gradients, and accordingly, they could be written symbolically as

$$U_0 = U_2 + U_3 + U_4 \tag{5.11}$$

The superscripts of the terms on the right-hand side of Eq. (5.11) denote the order of dependence of energy terms with respect to displacement gradients. To get the secant matrices in quadratic form in the additional displacement formulation (ADF), let us assume $\mathbf{g}^{\mathrm{T}} = \{u_{,x}\, u_{,y}\, v_{,x}\, v_{,y}\, w_{,x}\, w_{,y}\, w_{,xx}\, w_{,yy}\, w_{,xy}\}$ to be a vector which is the set of all displacement gradients of additional displacements found in Green's strain in Eq. (5.4) in which the comma represents the differentiation of the preceding quantity with succeeding indices. The gradients of initial imperfection 'w_0' being constants are excluded from the gradient vector $\mathbf{g}_{\alpha}$. However, in TDF (Eq. 5.2), the total displacement (i.e., $\overline{w} = w + w_0$) is a variable, and hence, the gradients of total displacements need to be considered in $\mathbf{g}$. It should be emphasized that the basis of the secant matrix technique (Mallet and Marcal 1968; Rajasekharan and Murray 1973; Onate 1995) is in expressing Eq. (5.10) in a quadratic form as

$$U_0 = \mathbf{g}^{\mathrm{T}} \mathbf{U} \mathbf{g} \tag{5.12}$$

which could be rewritten using Mallet and Marcal's (1968) symbolism as

$$U_0 = \mathbf{g}^{\mathrm{T}} \left[\frac{1}{2} \mathbf{K} + \frac{1}{2} \mathbf{N0} + \frac{1}{6} \mathbf{N1} + \frac{1}{12} \mathbf{N2} \right] \mathbf{g} \tag{5.13}$$

In the TDF, Eq. (5.2) is as assumed as the strain, and if it is applied in Eq. (5.10), the product could be written symbolically as

$$U_0 = \int_{\Omega} \left(\boldsymbol{\varepsilon}_1^1 + \boldsymbol{\varepsilon}_2^2 - \boldsymbol{\varepsilon}_3^0 - \boldsymbol{\varepsilon}_4^1\right)^{\mathrm{T}} C \left(\boldsymbol{\varepsilon}_1^1 + \boldsymbol{\varepsilon}_2^2 - \boldsymbol{\varepsilon}_3^0 - \boldsymbol{\varepsilon}_4^1\right) d\Omega \tag{5.14}$$

If we carry out the multiplication of terms in Eq. (5.14) and double differentiate to get a form as in Eq. (5.13), it can be shown that the product terms of that are of order less than quadratic, such as $\int \varepsilon_1^1 C \varepsilon_3^0 d\Omega$, $\int \varepsilon_3^0 C \varepsilon_1^1 d\Omega$, $\int \varepsilon_3^0 C \varepsilon_1^4 d\Omega$, $\int \varepsilon_4^1 C \varepsilon_3^0 d\Omega$, , would vanish. In other words, a part of the energy has not been included in the total potential energy formulation. It has been shown by Arul Jayachandran et al. (2001) that Yang's formulation or the total displacement formulation (TDF) is applicable only to small amounts of initial imperfection say 0.2 times thickness or less. When larger initial imperfections are present, those product terms may not become small enough to be neglected, producing erroneous results. On the other hand in the additional displacement formulation (ADF), Eq. (5.4) is used as Green's strain and the strain energy in Eq. (5.10) could be symbolically written as

$$U_0 = \int_{\Omega} \left(\boldsymbol{\varepsilon}_1^1 + \boldsymbol{\varepsilon}_2^2 - \boldsymbol{\varepsilon}_3^1 - \boldsymbol{\varepsilon}_4^1\right)^{\mathrm{T}} C \left(\boldsymbol{\varepsilon}_1^1 + \boldsymbol{\varepsilon}_2^2 - \boldsymbol{\varepsilon}_3^1 - \boldsymbol{\varepsilon}_4^1\right) d\Omega \tag{5.15}$$

Inspection of Eq. (5.15) reveals that every product term would result in a proper energy term in U_o and in double differentiation to get a form defined in Eq. (5.12), would contribute to the total potential energy. Hence, there is no approximation in the strain energy calculated. As defined in Eq. (5.13), the strain energy in terms of displacement gradients can be explicitly written before carrying out the numerical integration with respect to finite element coordinates. It is seen from Eq. (5.15) that six additional terms of strain energy vis-à-vis six additional incremental matrices has been accounted as compared with Eq. (5.14). Using Eq. (5.8), the additional energy terms which contribute to the total potential energy of the plate could be written as a function of displacement gradients as

$$\int \varepsilon_2^2 C \varepsilon_3^1 = k_1 w_{,x}^3 + k_2 w_{,y}^3 + k_3 w_{,x}^2 w_{,y} + k_1 w_{,y} w_{,x}^2 \tag{5.16}$$

where k_1, k_2, k_3, k_4 can be written as $k_1 = 0.5\ \mathrm{A}_{11}\ w_{0,x} + 0.5\mathrm{A}_{13}\ w_{0,y}$, $k_2 = 0.5\ \mathrm{A}_{22}\ w_{0,x} + 0.5\ \mathrm{A}_{23}\ w_{0,y}$, $k_3 = \mathrm{A}_{31}\ w_{0,x} + 0.5\ \mathrm{A}_{12}\ w_{0,y} + 0.5\ \mathrm{A}_{13}\ w_{0,x} + \mathrm{A}_{33}\ w_{0,y}$ and $k_4 = \mathrm{A}_{32}\ w_{0,y} + 0.5\ \mathrm{A}_{21}\ w_{0,x} + 0.5\ \mathrm{A}_{23}\ w_{0,y} + \mathrm{A}_{33}\ w_{0,x}$

$$\begin{aligned}\int \varepsilon_3^1 C \varepsilon_1^1 = {} & A_{11} w_{0,x} w_{,x} u_{,x} + A_{21} w_{0,y} w_{,y} u_{,x} + A_{31} w_{0,x} w_{,y} u_{,x} \\ & + A_{31} w_{0,y} w_{,x} u_{,x} + A_{12} w_{0,x} w_{,x} v_{,y} + A_{22} w_{0,y} w_{,y} v_{,y} \\ & + A_{32} w_{0,x} w_{,y} v_{,y} + A_{32} w_{0,y} w_{,x} v_{,y} + A_{13} w_{0,x} w_{,x} u_{,y} \\ & + A_{23} w_{0,y} w_{,y} u_{,y} + A_{33} w_{0,x} w_{,y} u_{,y} + A_{33} w_{0,y} w_{,x} u_{,y} \\ & + A_{13} w_{0,x} w_{,x} v_{,x} + A_{23} w_{0,y} w_{,y} v_{,x} + A_{33} w_{0,x} w_{,y} v_{,x} \\ & + A_{33} w_{0,y} w_{,x} v_{,x}.\end{aligned} \tag{5.17}$$

$$\int \varepsilon_1^1 C \varepsilon_1^1 = B_{11} w_{,xx} u_{,x} + B_{21} w_{,xx} v_{,y} + B_{31} w_{,xx} u_{,y} + B_{31} w_{,xx} v_{,x} + B_{12} w_{,yy} u_{,x}$$
$$+ B_{22} w_{,yy} v_{,y} + B_{32} w_{,yy} u_{,y} + B_{32} w_{,yy} v_{,x} + B_{13} w_{,xy} u_{,x}$$
$$+ B_{23} w_{,xy} v_{,y} + B_{33} w_{,xy} u_{,y} + B_{33} w_{,xy} v_{,x} \tag{5.18}$$

$$\int \varepsilon_2^2 C \varepsilon_1^1 = 0.5 B_{11} w_{,xx} w_{,x}^2 + 0.5 B_{21} w_{,xx} w_{,y}^2 + B_{31} w_{,xx} w_{,y} w_{,x}$$
$$+ 0.5 B_{12} w_{,yy} w_{,x}^2 + 0.5 B_{22} w_{,yy} w_{,y}^2 + B_{32} w_{,yy} w_{,y} w_{,x}$$
$$+ B_{13} w_{,xy} w_{,x}^2 + B_{23} w_{,xy} w_{,y}^2 + 2 B_{33} w_{,xy} w_{,x} w_{,y}. \tag{5.19}$$

$$\int \varepsilon_3^1 C \varepsilon_1^1 = B_{11} w_{0,x} w_{,xx} w_{,x} + B_{21} w_{0,y} w_{,xx} w_{,y} + B_{31} w_{0,x} w_{,xx} w_{,y}$$
$$+ B_{31} w_{0,y} w_{,xx} w_{,x} + B_{12} w_{0,x} w_{,yy} w_{,x} + B_{22} w_{0,y} w_{,yy} w_{,y}$$
$$+ B_{32} w_{0,x} w_{,yy} w_{,y} + B_{32} w_{0,y} w_{,yy} w_{,x} + 2 B_{13} w_{0,x} w_{,xy} w_{,x}$$
$$+ 2 B_{23} w_{0,y} w_{,xy} w_{,y} + 2 B_{33} w_{0,y} w_{,xy} w_{,y} + 2 B_{33} w_{0,y} w_{,xy} w_{,x} \tag{5.20}$$

$$\int \varepsilon_1^1 C \varepsilon_1^1 = D_{11} w_{,xx}^2 + D_{21} w_{,xx} w_{,yy} + 2 D_{31} w_{,xx} w_{,xy} + D_{12} w_{,xx} w_{,yy} + D_{22} w_{,yy}^2$$
$$+ D_{32} w_{,yy} w_{,xy} + 2 D_{13} w_{,xx} w_{,xy} + 2 D_{23} w_{,xy} w_{,yy} + 2 D_{33} w_{,xy}^2 \tag{5.21}$$

The inadequacies of Yang's formulation (1971) (TDF) were noticed while computing the approximate collapse loads of thin plates (Arul Jayachandran et al. 1995). Hence in the present study, Eq. (5.4) is taken as Green's strain of an initially imperfect plate and the complete set of displacement gradients for the secant matrices for the postbuckling analysis of thin initially imperfect laminated composite plates have been derived and are presented in the Appendix of Chap. 6.

As far as the explicit matrices are concerned, it is seen from the literature that Mallet and Marcal (1968) first presented explicit coefficients of **K, N1, N2** for a perfectly flat plate element, and these were later pointed out to be inadequate by Rajasekaran and Murray (1973). This book presents the correct and improved secant matrices for the nonlinear analysis of initially imperfect laminated composite plates. For the finite element implementation, the potential of the laminated composite plate, as given in Eq. (5.10), can be represented as the sum of potentials contributed by individual plate finite elements. The element displacements $\mathbf{p}_e$ can be interpolated using a linear combination of the generalized nodal displacements $\mathbf{p}$ and an approximating shape function $\boldsymbol{\varphi}$ as

$$\mathbf{p}_e = \varphi \mathbf{p}. \tag{5.22}$$

The displacement gradients can be related to the generalized nodal displacements of a typical plate element as

$$\mathbf{g} = \mathbf{D}\,\mathbf{p} \tag{5.23}$$

D is a matrix that relates the finite element's displacement gradients and nodal displacements. Using Eq. (5.13), the strain energy equation can be written symbolically as

$$U_0 = \mathbf{p}^{\mathrm{T}}\left(\mathbf{D}^{\mathrm{T}}\mathbf{U}\mathbf{D}\right)\mathbf{p}. \tag{5.24}$$

The total potential energy can be expressed symbolically in terms of generalized nodal displacements **p** as

$$\pi^e = \frac{1}{2}\mathbf{p}^{\mathrm{T}}\mathbf{K}\mathbf{p} + \frac{1}{2}\mathbf{p}^{\mathrm{T}}\mathbf{N0}\mathbf{p} + \frac{1}{6}\mathbf{p}^{\mathrm{T}}\mathbf{N1}\mathbf{p} + \frac{1}{12}\mathbf{p}^{\mathrm{T}}\mathbf{N2}\mathbf{p} - \mathbf{p}^{\mathrm{T}}\mathbf{q} \tag{5.25}$$

It should be noted that the deformation of the initially imperfect plate has been measured with respect to the additional deflection as defined in Eq. (5.4), and hence, all the incremental matrices in Eq. (5.25) would be functions of the additional displacement vector $\boldsymbol{p}$.

5.4 Equilibrium Equations and Analysis Method

The equilibrium equation is derived by applying the equilibrium condition to Eq. (5.25), resulting in $\partial\pi/\partial\mathbf{p} = 0$ which can be expressed as:

$$\mathbf{q} = \mathbf{K}\mathbf{p} + \mathbf{N0}\mathbf{p} + \frac{1}{2}\mathbf{N1}\mathbf{p} + \frac{1}{3}\mathbf{N2}\mathbf{p} \tag{5.26}$$

As previously mentioned, the secant matrices in this equation are solvable through an iterative process. However, by further differentiating the equation with respect to **p**, the tangent stiffness matrix relation is obtained as:

$$\Delta\mathbf{q} = \mathbf{K}\Delta\mathbf{p} + \mathbf{N0}\Delta\mathbf{p} + \mathbf{N1}\Delta\mathbf{p} + \mathbf{N2}\Delta\mathbf{p} \tag{5.27}$$

where Δ represents the incremental operator. In this study, cubic Hermitian polynomials are assumed for in-plane displacements u (with C^2 continuity) and out-of-plane displacements w (also with C^2 continuity), as proposed by Bogner et al. (1966). Despite numerous developments that had taken place in the evolution of the new plate bending elements, the Bogner et al. (1966) element has been used because the element yields very good results in the case of postbuckling problems (Harris and Little 1991). The equilibrium equation is solved incrementally by first computing the Euler predictor $\Delta\mathbf{p}$ corresponding to a load increment $\Delta\mathbf{q}$ and then correcting with Newton corrector iterations by computing the unbalanced forces from the equilibrium equations. For problems involving snap-through behavior, the 'arc length' and

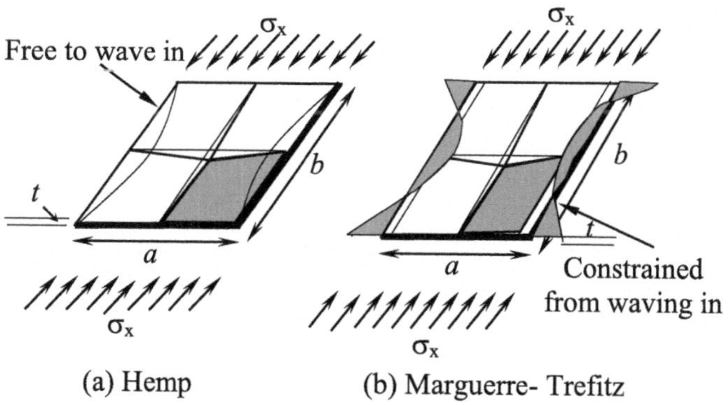

Fig. 5.3 Membrane boundary conditions encountered in axially loaded plates

the minimum residual displacement method procedure have been implemented in the present study. In the case of postbuckling analysis of thin plates under axial compression, non-classical kinematic boundary conditions such as unloaded edges as free to wave in or moving in straight, while the loaded edges remaining straight during axial deformation are encountered. In the literature, these two boundary conditions are named Hemp and Marguerre-Trefitz boundary conditions which are illustrated in Fig. 5.3.

The present study manages these boundary conditions using the Lagrange multiplier technique. While this method disrupts the banded structure of the stiffness matrix, specialized bookkeeping algorithms have been developed to retain the banded format of both the tangent and secant matrices through modified skyline storage.

5.5 Numerical Studies

Studies were carried out to assess the performance of the numerical formulations derived in this study. First, the large deflection behavior of simply supported isotropic plate subjected to uniform pressure load was studied, for two different initial imperfections of $w_0/t = 0.1$ and $w_0/t = 1.0$. The analytical results of Rushton (1970) which are claimed to be accurate to within 0.5% of error, were taken as the basis for comparison. The large deflection behavior was investigated in both cases, and the results are presented in Table 5.1. It is seen from Table 5.1, that the results of the present study and the results of Yang's formulation are quite accurate in the case of initial imperfection $w_0/t = 0.1$. Both the displacements and stresses are seen to be in good agreement for this case. Nevertheless, for the case of $w_0/t = 1.0$, the results of displacements and stresses of Yang's formulation are highly erroneous in the initial stages of the loading, while the results of the present study compare very well with the results of Rushton (1970).

Table 5.1 Comparison of results of large deflection of isotropic plates with initial curvature

Simply supported square plate—initial imperfection $w_0/t = 0.1$ membrane displacements immovable						
Pressure load $\frac{qb^4}{Et^4}$	Out-of-plane center deflections w/t			Stresses at the unloaded face at the center $\frac{\sigma b^2}{Et^2}$		
	Rushton (1970)	Yang's formulation	Equation (5.26) and Eq. (5.27)	Rushton (1970)	Yang's formulation	Present study
9.16	0.316	0.314	0.313	2.46	2.482	2.452
36.6	0.747	0.745	0.744	6.51	6.498	6.477
146.5	1.38	1.376	1.376	15.0	13.88	13.860
Simply supported square plate—initial imperfection $w_0/t = 1.0$ membrane displacements immovable						
9.16	0.089	0.180	0.092	0.94	5.382	0.933
36.60	0.289	0.354	0.292	3.16	6.232	3.138
146.50	0.730	0.767	0.738	8.86	10.702	8.852

For the postbuckling analysis of thin plates, two initial imperfections were considered, namely $w_0/t = 0.1$ and $w_0/t = 0.5$. The results are presented in Fig. 5.4. It is seen from Fig. 5.4 that the results of Yang's formulation and the present study compare very well with the results of Rhodes (1986) for an imperfection of $w_0/t = 0.1$. However, for an imperfection of $w_0/t = 0.5$, the results of Yang's formulation are erroneous (Fig. 5.4). The results of the present study present an excellent comparison with the results of Rhodes (1986) both for imperfections $w_0/t = 0.1$ and $w_0/t = 0.5$.

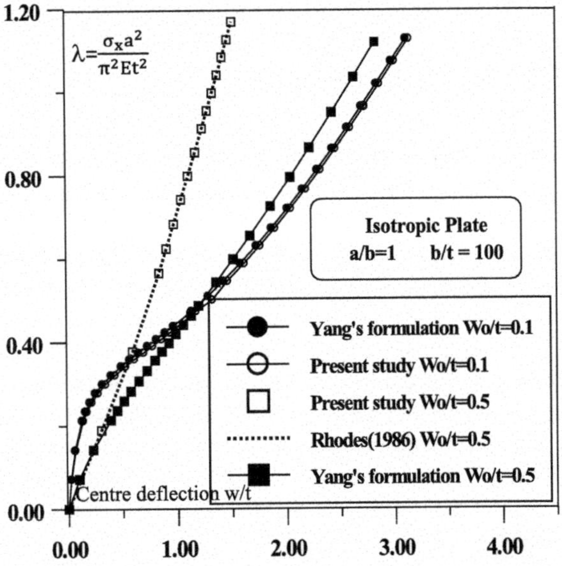

Fig. 5.4 Comparison of Yang's formulation and the present formulation

The collapse predictions of thin plates under compression were carried out for both free and constrained wave in membrane boundary conditions. Four initial imperfection amplitudes were chosen, and the collapse loads σ_{av}/σ_y (average stress/yield stress) were calculated. For the Hemp boundary condition, the collapse was assumed to have reached when the membrane stress anywhere along the unloaded edges reached yield stress.

For the Marguerre-Trefftz boundary condition, collapse is assumed to have occurred when the combined stress $\sigma_x + 0.6\sigma_y$ reaches yield value. The results of the present study using the additional displacement formulation (ADF), shown in Table 5.2, are compared with the total displacement formulation (TDF) of Yang(1971) and the elastoplastic finite element solutions of Harris and Little(1991), wherein the bracketed terms represent % error. Table 5.2 shows that the results of the present study show much-improved accuracy over the total displacement formulation compared with an elaborate elastoplastic finite element analysis of Harris and Little (1991). It is seen from Table 5.2 that as the initial imperfection is increased, the error involved in the TDF increases. Hence, it has been demonstrated that the improved secant matrices are very accurate for the collapse prediction of thin plates.

A study was carried out using the secant matrices developed in this chapter to assess the reason for this discrepancy between the TDF and ADF. For a large deflection thin imperfect plate problem, the strain energy contributed by the terms such as linear, initial imperfection, first order, and second order was carried out, and the results are presented in Figs. 5.5 and 5.6 for initial imperfection $w_0/t = 0.1$ and $w_0/t = 1.0$. It is seen from Fig. 5.5 that for the imperfection of $w_0/t = 0.1$, the contribution of strain energy by the initial imperfection terms is much less almost throughout the loading.

But in contrast, in Fig. 5.5 it is seen that in the initial stages of loading, the contribution to the strain energy by the initial imperfection terms is significant. Hence, it is inferred that the approximation of strain energy involved in Yang's formulation

Table 5.2 Collapse loads of simply supported thin isotropic plates under compression

	Membrane boundary conditions					
	Free to wave in (Hemp)			Constrained to wave in straight (Margurre-Trefftz)		
	σ_{av}/σ_y			σ_{av}/σ_y		
w_0/t	Yang's Formulation TDF	Equation (5.26) and Eq. (5.27) ADF	Harris and Little (1991)	Yang's formulation TDF	Equation (5.26) and Eq. (5.27) ADF	Harris and Little (1991)
0.005	0.575 (1.5)	0.578 (1.0)	0.584	0.652 (1.3)	0.653 (1.2)	0.661
0.050	0.583 (0.5)	0.577 (0.5)	0.580	0.624 (5.8)	0.651 (0.7)	0.656
0.150	0.552 (3.4)	0.573 (0.5)	0.570	0.614 (5.1)	0.647 (0.0)	0.647
0.350	0.525 (5.0)	0.564 (2.1)	0.552	0.586 (7.3)	0.639 (1.4)	0.630

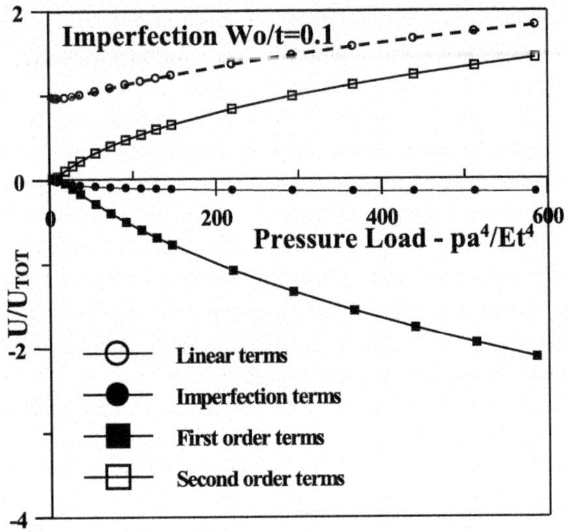

Fig. 5.5 Strain energy contributed by different terms—large deflection bending of simply supported plate

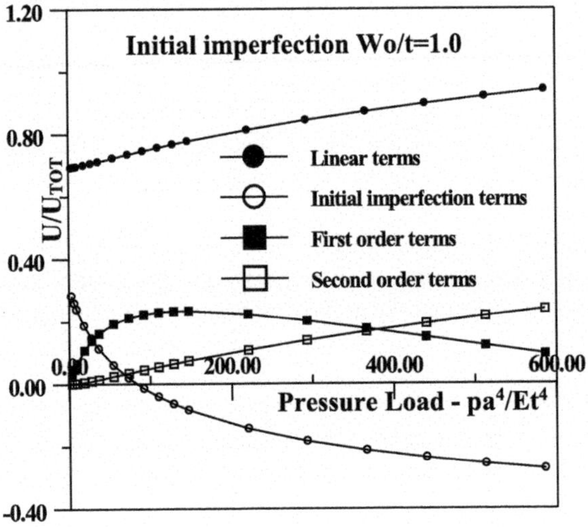

Fig. 5.6 Strain energy contributed by different terms—large deflection bending of simply supported plate

(TDF) is not justified for larger imperfection values greater than $w_0/t = 0.2$ (Arul Jayachandran et al. 2001).

The modeling of imperfection is very important because the initial postbuckling behavior is very important in the design of metal and laminated composite plates. This is probably one of the reasons for the inaccuracies found in the results of TDF in Table 5.1.

As a next problem, postbuckling analysis of a boron-epoxy plate ($E_T/E_L = 0.1$, $G_{LT}/E_L = 0.0333$, $\nu_{LT} = 0.22$) was carried out with an initial imperfection of w_0/t

= 0.0. The results are presented in Table 5.3. The results are in excellent agreement with the analytical results of Prabhakara and Chia (1973).

Figure 5.7 shows the results of the postbuckling analysis of boron-epoxy plate ($E_T/E_L = 0.1, G_{LT}/E_L = 0.0333, \nu_{LT} = 0.22$) with three initial imperfections, namely $w_0/t = 0.0$, $w_0/t = 0.1$, and $w_0/t = 1.0$. For an initial imperfection of $w_0/t = 0.0$, a branching from the primary path to the secondary path is observed, as expected.

To highlight the advantage of the proposed improved secant matrices presented in this study, the postbuckling analysis of an imperfect isotropic plate with both local and global imperfections was carried out. An imperfection of $w_0/t = 0.1$ was assumed to be present globally and in another case dent like circular local imperfection of

Table 5.3 Comparison of results for the postbuckling analysis of boron-epoxy plate under uniaxial compression

Simply supported Boron-Epoxy square plate – initial imperfection $w_0/t = 0.0$ membrane displacements allowed			
	Out-of-plane center deflections w/t		
P/P_{cr}	Prabhakara and Chia (1973)	Equation (5.26) and Eq. (5.27)	% Difference
1.00	0.00	0.01	1.15
1.25	1.85	1.82	1.35
1.50	2.68	2.64	1.36
2.00	3.95	3.95	1.36
2.50	5.03	5.07	0.79
3.00	6.03	6.13	1.67

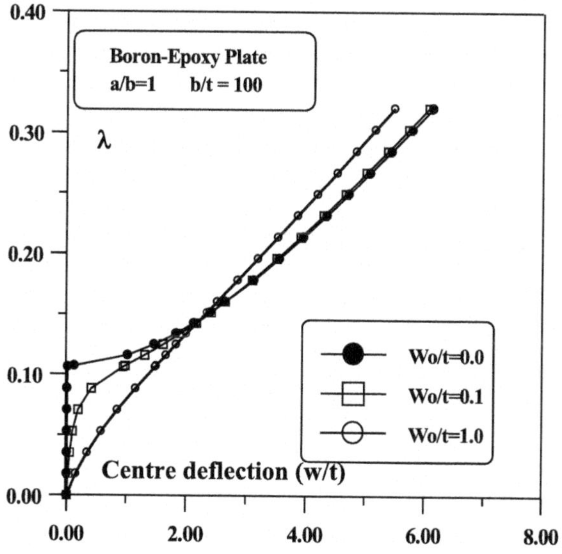

Fig. 5.7 Postbuckling paths of simply supported boron-epoxy plate under uniaxial compression

Fig. 5.8 Postbuckling paths of simply supported isotropic plate—effect of type of imperfection

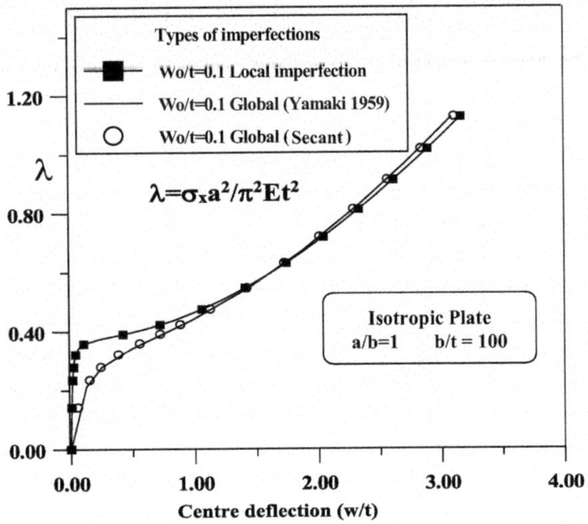

diameter '0.25a' with an imperfection of $w_0/t = 0.1$ was assumed at the center of the plate. The results of the study are presented in Fig. 5.8. It is interesting to note that the postbuckling behavior of the plate with localized dent like imperfection is qualitatively different from the globally imperfect plate. The former is much closer to the bifurcation type of buckling observed in perfect plates, and the difference is quite high near the linear critical load or in the initial postbuckling stages.

This example also underlines the importance of initial postbuckling behavior for the design of laminated composite plates and the necessity of considering localized imperfection and random imperfections found in the case of plates.

An interesting part of the postbuckling analysis of thin plates, especially in the case of uniaxially compressed plates is that, the membrane or the in-plane boundary condition characterizes the postbuckling behavior. These two boundary conditions explained in Sect. 5.4 were referred to as 'movable' and 'immovable' by Librescu et al. (1993) and the details of the plate analyzed are given in Fig. 5.9 and Poisson's ratio ν_{LT} value was chosen as 0.25. The postbuckling analyses of plates for these two boundary conditions are carried out, and the results are presented in Fig. 5.9. The results are in good agreement with the analytical results of Librescu et al. (1993) for laminated composite plates. Since the results obtained in the present study almost traced the values presented by Liberscu et al. (1993), results of the present study alone are presented. For the isotropic plate, postbuckling studies were carried out for the Hemp boundary condition and the results of the present study are compared with the analytical results of Yamaki (1959) in Table 5.4. Another example of a cross-ply plate (90/0) ($E_T/E_L = 0.1$, $G_{LT}/E_L = 0.025$, $\nu_{LT} = 0.3$) was solved, and the results are presented in Fig. 5.10. The results are in good agreement with the results of Kim et al. (1998). This is a classic example of a perfect plate undergoing a non-bifurcation type of buckling due to asymmetric ply orientation. In this chapter, it has been demonstrated that the secant matrices have been found to be very accurate for the postbuckling analysis of isotropic and laminated composite plates.

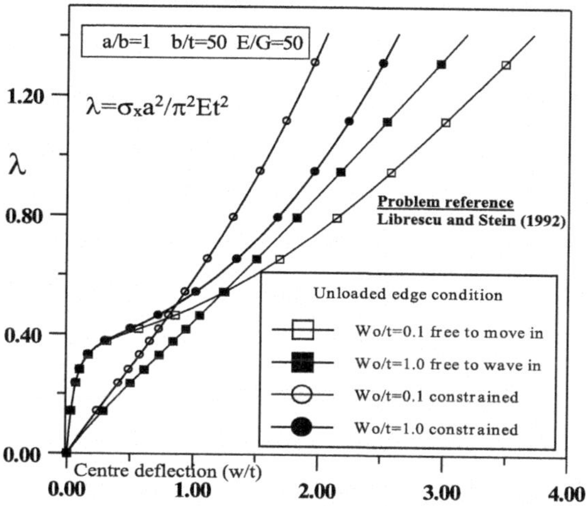

Fig. 5.9 Postbuckling paths of a single layer, simply supported composite plate

Table 5.4 Comparison of results postbuckling analysis of isotropic plate under uniaxial compression

Simply supported isotropic square plate – initial imperfection $w_0/t = 0.1$ – unloaded edges free to wave in (movable) – 2 × 2 mesh for one-quarter plate

Axial load $\lambda = \sigma_x a^2/\pi^2 Et^2$	Out-of-plane center deflections w/t	
	Yamaki (1959)	Equation (5.26) and Eq. (5.27)
0.212	0.124	0.123
0.278	0.249	0.244
0.342	0.495	0.483
0.390	0.738	0.724
0.438	0.976	0.963
0.546	1.434	1.423
0.672	1.870	1.865
0.812	2.288	2.288
0.963	2.692	2.698
1.126	3.090	3.111

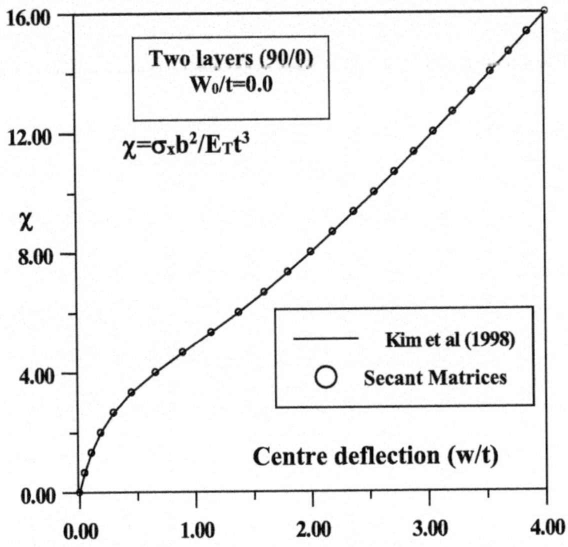

Fig. 5.10 Postbuckling path of simply supported boron/epoxy cross-ply laminated plate

5.6 Summary

The importance of considering the initial imperfections and their effect on the postbuckling behavior of laminated composite and isotropic plates has been elaborated. In deriving the incremental matrices using the secant matrix technique for the postbuckling range of laminated composite plates, it has been found that the formulation presented in the literature in terms of total displacement $(w + w_0)$ is erroneous. The strain energy of certain lower order terms vanishes in the secant matrices formed by double differentiation using total displacements. This approximation is justified only in the case of small imperfections say less than $w_0/t = 0.2$. Since practically observed imperfections are of the order of $w_0/t = 1.0$ their utility is limited.

In the present study, secant matrices are presented for the postbuckling analysis of thin laminated composite plates (CLPT) with initial imperfection, which are found to be accurate for all practical range of imperfections. Using the developed incremental matrices, several studies on postbuckling analysis of laminated composite plates based on the CLPT and isotropic plates have been carried out.

References

Aravind HB (1990) Nonlinear analysis of isotropic and orthotropic plated members using finite strips, Ph.D thesis, IIT, Madras, India

Arul Jayachandran S, Gopalakrishnan S, Narayanan R (2001) Incremental matrices for the post-buckling analysis of thin plates with small initial curvature. Int J Struct Eng Mech 12:283–295. https://doi.org/10.12989/sem.2001.12.3.283

Arul Jayachandran S, Vaidyanathan CV, Rao TVSRA (1995) Prediction of collapse loads of thin plates under compression. In: Rajasekharan S, Sridharan S (ed) Coimbatore, India, pp 561–575

Arthur W Leissa (1987) A review of laminated composite plate buckling. App Mech Rev 40:575–590

Atluri SN (1997) Structural integrity and durability, Tech. Science Press, Forsyth, GA

Balamurgan V, Ganapathy M, Varadhan TK (1996) Nonlinear dynamic instability of laminated composite plates using finite element method. Comp Struct 60:125–130. https://doi.org/10.1016/0045-7949(95)00368-1

Barbero EJ, Reddy JN (1990) Nonlinear analysis of composite laminates using a generalized laminated plate theory. AIAA J 28:1987–1995. https://doi.org/10.2514/3.10509

Bradfield CD (1982) An evaluation of the elastic-plastic analysis of steel plates loaded by uniaxial in-plane compression. Int J Mech Sci 24:127–146. https://doi.org/10.1016/0020-7403(82)90033-9

Bogner FK, Fox RL, Schmidt LA (1966) The generation of inter – element compatible matrices by the use of interpolation formulas. Proceedings of the conference on matrix methods in structural mechanics, TR 66-80, Air Force Flight Dynamics Lab., Wright– Patterson Airforce Base, Ohio. 397–443

Chang B, Shabana AA (1990) Total lagrangian formulation for the large displacement analysis of rectangular plates. Int J Num Meth Eng 24:73–103. https://doi.org/10.1002/nme.1620290107

Chang B, Sherboune (1990) Nonlinear finite element formulation for the large displacement analysis of plates. J Appl Mech 57:707–718. https://doi.org/10.1115/1.2897081

Chia CY (1988) Geometrically nonlinear behavior of composite plate: a review. App Mech Rev 41:439–450. https://doi.org/10.1115/1.3151873

Datoo MH (1991) Mechanics of fibrous composites. Elsevier App Sci Co

Dawe DJ, Lam SSE, Azizian ZG (1993) Finite strip post local buckling analysis of composite prismatic plate structures. Comput Struct 48:1011–1123. https://doi.org/10.1016/0045-7949(93)90436-H

Dow RS, Smith RS (1984) Effects of localized imperfection on compressive strength of long rectangular plate. J Cons Steel Res 4:51–76. https://doi.org/10.1016/0143-974X(84)90035-X

Ganapathy M, Varadhan TK (1995) Nonlinear free flexural vibrations of laminated circular cylindrical shells. Compos Struct 30:33–49. https://doi.org/10.1016/0263-8223(94)00025-5

Harris JP, Little GH (1991) Aspects of the behavior of rectangular steel plates under uniaxial compression. Struct Eng 69:255–261

Kapania RK, Yang TY (1987) Buckling postbuckling and nonlinear vibration of imperfect plates. AIAA J 25:1338–1346

Kim KD, Park T, Voyiadjis GZ (1998) Postbuckling analysis of composite panels with imperfection damage. Comput Mech 22:375–387. https://doi.org/10.1007/s004660050369

Kumar A, Kishore BNR (1991) Buckling of anti-symmetric angle and cross-ply rectangular plates under shear and compression. Int J Mech Sci 33:31–39. https://doi.org/10.1016/0020-7403(91)90025-X

Lam SSE, Dawe DJ (1998) Nonlinear analysis of rectangular laminated plates under end shortening using finite strip method. In: Leondes CT, Gordon

Librescu L, Souza MA (1993) Postbuckling of geometrically imperfect shear deformable panels under combined thermal and compressive edge loading. J App Mech Trans ASME 60:526–533. https://doi.org/10.1115/1.2900825

Lui TH, Lam SSE (2001) Finite strip analysis of laminated plates with general initial imperfection under end shortening. Eng Struct 23:673–686. https://doi.org/10.1016/S0141-0296(00)00076-6

Madasamy CM, Kalyanaraman V (1994) Analysis of plated structures with rectangular cut-outs and internal supports using the spline finite strip method. Comput Struct 52:277–286. https://doi.org/10.1016/0045-7949(94)90280-1

Mallet RH, Marcal PV (1968) Finite element analysis of nonlinear structures. J Struct Div ASCE 94:2081–2105. https://doi.org/10.1061/JSDEAG.0002066

Narayanan G, Krishnamoorthy CS (1989) A simplified formulation for a plate/shell element for geometrically nonlinear analysis of moderately large deflections with small rotations. Finite Elem Anal des 4:333–350. https://doi.org/10.1016/0168-874X(89)90027-9

Noor AK (1995) Finite element buckling and postbuckling analysis. In: Turvey GJ, Marshall IH (eds) buckling and postbuckling analysis of composite plates. Chapman & Hall

Onate E (1995) On the derivation and possibilities of the secant stiffness matrix for nonlinear finite element analysis. Comp Mech 15:572–593. https://doi.org/10.1007/BF00350269

Paik JK (1995) A new concept of the effective shear modulus for a plate buckled in shear. J Ship Res 39:70–75. https://doi.org/10.5957/jsr.1995.39.1.70

Prabhakara MK, Chia CY (1973) Postbuckling behavior of rectangular orthotropic plates. J Mech Eng Sci 15:25–33. https://doi.org/10.1243/JMES_JOUR_1973_015_006_02

Rajasekaran S, Murray DW (1973) Incremental finite element matrices. J Struct Eng ASCE 99:2423–2438. https://doi.org/10.1061/JSDEAG.0003667

Rhodes J (1986) Micro computer design analysis of plate postbuckling behavior. J Strain Anal 21:71–76. https://doi.org/10.1243/03093247V212071

Rushton KR (1970) Large deflection of plates with initial curvature. Int J Mech Sci 12:1037–1051. https://doi.org/10.1016/0020-7403(70)90031-7

Saigal S, Kapania RK, Yang TY (1986) Geometrically nonlinear finite element analysis of imperfect laminated shells. J Comp Mat 20:197–213. https://doi.org/10.1177/002199838602000206

Schafer BW, Pekoz T (1998) Computational modelling of cold formed steel: charaterisation of imperfections and residual stresses. J Const Steel Res 47:193–210. https://doi.org/10.1016/S0143-974X(98)00007-8

Shanmugam NE (1987) Strength of axially loaded orthotropic plates. J Struct Eng ASCE 113:3322–3333. https://doi.org/10.1061/(ASCE)0733-9445(1987)113:2(322)

Shen HS (1997) Thermal postbuckling analysis of imperfect laminated plates using a higher order shear deformation theory. J App Mech ASME 32:1035–1050. https://doi.org/10.1016/S0020-7462(96)00135-7

Shen HS, Williams FW (1996) Postbuckling analysis of imperfect laminated plates under combined axial and thermal loads. Comput Mech 17:226–233. https://doi.org/10.1007/BF00364825

Shin DK, Griffin JROH, Gurdal Z (1993) Postbuckling response of laminated plates under uniaxial compression. Int J Nonlin Mech 28:95–115. https://doi.org/10.1016/0020-7462(93)90009-A

Singh G, Rao GV, Iyengar NGR (1994) Thermal postbuckling behavior of laminated composite plates. AIAA J 32:1336–1338

Sridharan Srinivasan (1982) A finite strip analysis of locally buckled plated structures subjected to non-uniform compression. Engineering Structures, 4:249–255

Sundaresan P, Singh G, Rao GV (1996) Buckling and postbuckling analysis of moderately thick laminated rectangular plates. Comput Struct 61:79–86. https://doi.org/10.1016/0045-7949(96)00010-7

Ueda Y, Yao T (1985) The influence of complex initial deflection modes on the behavior and ultimate strength of rectangular plates in compression. J Const Steel Res 5:265–302. https://doi.org/10.1016/0143-974X(85)90024-0

Usami T (1993) Effective width of locally buckled plates in compression and bending. J Struct Eng ASCE 119:1358–1373. https://doi.org/10.1061/(ASCE)0733-9445(1993)119:5(1358

Yamaki N (1959) Postbuckling behavior of rectangular plates with small initial curvature loaded in edge compression. J App Mech 26:407–415. https://doi.org/10.1115/1.4012053

Yang TY (1971) A finite element procedure for large deflection analysis of plates with initial deflections. AIAA J 9:1468–1473

Yang TY, Han AD (1983) Buckled plate vibrations and large amplitude vibration using higher order triangular elements. AIAA J 21:758–766. https://doi.org/10.1007/978-3-322-89564-6_13
Yang TY (1972) Elastic snap through analysis of curved plates using discrete elements. AIAA J 10(4):371–372

Chapter 6
Postbuckling Analysis of Shallow Shells Using Secant Matrices Based on Marguerre's Shell Theory

Abstract This chapter deals with the postbuckling analysis of shallow shells using secant matrices based on Marguerre's shell theory. Thin shell structures are common in civil, aerospace, and marine fields and are particularly susceptible to elastic buckling due to their slenderness. The previous chapter dealt with secant matrices for the postbuckling analysis of initially imperfect thin plates. This chapter extends Marguerre's shell theory into the nonlinear postbuckling domain by incorporating the initial imperfections as the shape of the shallow shell. Hence, we see that the ADF formulation presented in Chap. 5 is accurate enough that it also applies to shallow shells with the shape of the shell modeled as an imperfection from the flat geometry. Several examples of shallow shells under transverse and axial loads are presented in this chapter, validating ADF formulation. Results illustrate the effectiveness of the secant matrix method in capturing the complex postbuckling response of shallow shells.

Keywords Shallow shells · Secant matrix · Marguerre's shell theory · Postbuckling analysis · Thin-walled structures · Imperfection sensitivity · Nonlinear shell behavior · Finite element method · Composite shells · Structural stability

6.1 General

Thin shell structures are widely used in civil, aerospace, and marine structures due to their high strength-to-weight ratio. Nevertheless, their thinness increases the problem of elastic buckling under compressive loads, and a significant amount of research has been carried out to establish safe design criteria for practical shells. But this effort continues, as seen from the research work reported in the literature even in recent times.

A schematic diagram of a shell panel where its ordinates are modeled as a plate with imperfections is shown in Fig. 6.1. The buckling and postbuckling behavior of such shell panels have several distinct features compared to the postbuckling behavior

A. Jayachandran, *Elastic Postbuckling Analysis of Structural Systems*, Solid Mechanics and Its Applications 232, https://doi.org/10.1007/978-3-031-89936-2_6

of thin plates. Many shell loading problems exhibit small displacement, prebuckling behavior shown by the line OA in the schematic load vs displacement response curve shown in Fig. 6.2. Many theoretical studies aimed at determining the point B on the line OA, which is the point at which perfect shell would begin to buckle. In other words, it is called the elastic critical load of the shell structure. These analyses are basically linear eigenvalue analyses, providing the shell's buckling modes but giving no information about their subsequent stability. Only a study concerning 'postbuckled behavior' concentrates on the nonlinear problem of determining the large displacement behavior of shells with plausible assumptions of postbuckled modes. Many common shells possess limit loads considerably lesser than the classical load (e.g., curve DEF in Fig. 6.2), casting doubt on the suitability of using this 'linear critical load' as a basis for design. In addition to this, shells can be highly sensitive to geometry and imperfections. The curve OGH in Fig. 6.2 represents the load–deflection paths of imperfect shells. The shell carries the load until it reaches the local maximum at G, where it snaps to a highly buckled state at point H.

Naturally, the question is how much reduction in the load occurs for a given size of imperfection. This can be partially answered by determining the slope of the initial postbuckling path at B. A rising line BJ signifies that the shell is insensitive to imperfections. On the other hand, a falling line BC is imperfection-sensitive. However, a

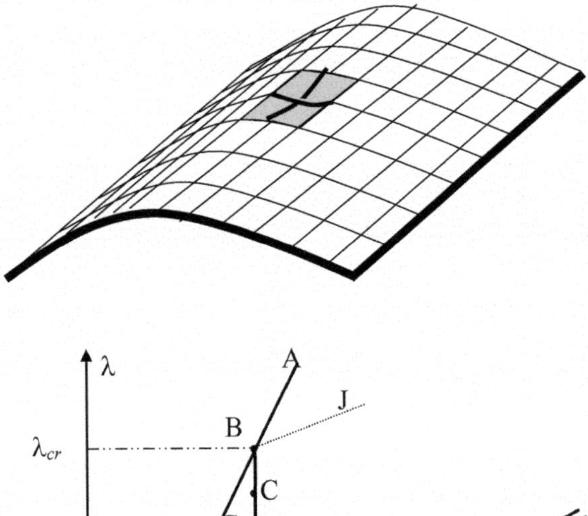

Fig. 6.1 Curved panel modeled as a plate with initial imperfection

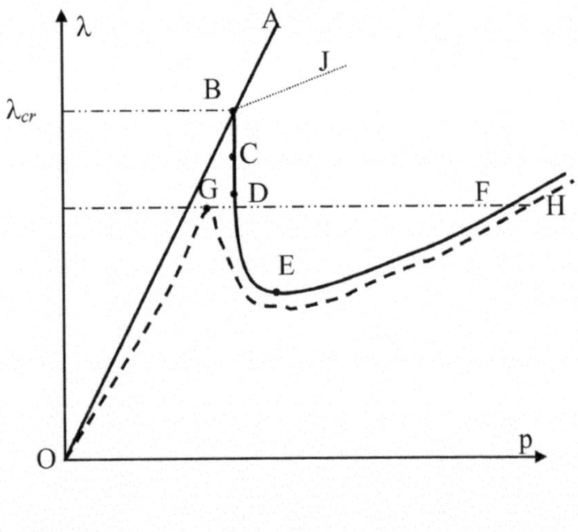

Fig. 6.2 Schematic representation of shell buckling phenomenon

detailed postbuckling analysis reveals the complete behavior of the shell. The above discussion underlines that the imperfection sensitivity of a shell is paramount in its design. Since imperfections are random in practice, correlating the buckling strength of imperfect shells to the statistical description of the initial imperfection may be a better approach. Such an approach is beyond the scope of this book.

6.2 Finite Element Analysis of Shell Structures

The primary consideration in the finite element analysis of shell structures is modeling the geometry of the curved shell surfaces. A large number of papers have been published on finite element analysis of shells, and they can be mainly grouped into four categories (Geis 1977): those using,

- Flat elements.
- Curved elements based on shallow shell theory.
- Curved elements based on deep shell theory.
- Solid elements.

The special finite elements for a particular geometrical shape, such as axisymmetric and cylindrical shells, fall into another category.

6.2.1 Shell Geometry Representation Through Flat Elements

The most straightforward approach to modeling the curved shell surface is an assemblage of flat-faceted finite elements. Replacing the curved shell surface with a finite number of flat elements introduces additional geometric errors in the finite element approximation. This error could be minimized by increasing the number of elements to model the geometry. Nevertheless, slope discontinuities between adjacent elements may produce local bending moments in shell regions when no bending occurs. Several papers have been published in the literature, such as elements with drilling degrees of freedom, elements with fictitious springs, and elements with average out-of-plane rotation to alleviate the problems encountered in using flat-faceted elements.

6.2.2 Curved Elements Based on Shallow Shell Theory

The principle of shallow shell theory lies in introducing the shallow shell terms to the governing strain–displacement relation based on certain assumptions. The present formulation falls into this category, and hence, the derivation of the shallow shell theory is briefly discussed below. Consider a shell segment represented by a curvilinear coordinate θ_1, θ_2, θ_3 referenced from a fixed Cartesian coordinate x, y, z

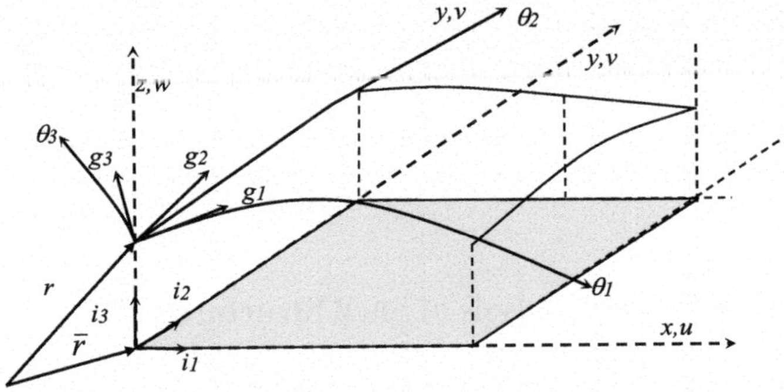

Fig. 6.3 Shell coordinates for Marguerre's shell theory

as shown in Fig. 6.3. It is assumed that the layers in the shell remain elastic during deformation, the generalized Hooke's law is valid, and no-slip occurs between any two layers. The shell surface is projected onto the x–y plane illustrated in Fig. 6.3. The position vector **r** of the reference point on the shell surface can be written as

$$\mathbf{r} = \bar{\mathbf{r}} + z\mathbf{i}_3 \tag{6.1}$$

where $\bar{\mathbf{r}}$ is the position vector of the origin of reference. The distance z between the x, y plane and the reference surface can be specified as a function of the coordinates x and y as

$$z = z(x, y). \tag{6.2}$$

Now the position vector '**r**' can be written concerning the x, y, z plane by replacing the curvilinear coordinates θ_1, θ_2 by the Cartesian coordinate x, y as

$$\mathbf{r} = x\mathbf{i}_1 + y\mathbf{i}_2 + z(x, y)\mathbf{i}_3. \tag{6.3}$$

Hence, the base vectors of the shell surface '**g**' get transformed as

$$\begin{aligned}
\mathbf{g}_1 &= \frac{\partial \mathbf{r}}{\partial x} = \mathbf{i}_1 + z_{,x}\mathbf{i}_3 \\
\mathbf{g}_2 &= \frac{\partial \mathbf{r}}{\partial y} = \mathbf{i}_2 + z_{,y}\mathbf{i}_3 \\
\mathbf{g}_3 &= \frac{\mathbf{g}_1 \times \mathbf{g}_2}{|\mathbf{g}_1 \times \mathbf{g}_2|} = -z_{,x}\mathbf{i}_1 + z_{,y}\mathbf{i}_2 + \mathbf{i}_3
\end{aligned} \tag{6.4}$$

The curvature and the torsion of the reference surface can be shown to be

$$\frac{1}{R_x} = w_{,xx} \quad \frac{1}{R_y} = w_{,yy} \quad \frac{1}{R_{xy}} = w_{,xy}. \tag{6.5}$$

Using Eqs. (6.4) and (6.5), the membrane strains ε_x, ε_y, ε_{xy} and the bending strains χ_x, χ_y, χ_{xy} can be written (Vlasov 1957) as

$$\begin{aligned} \varepsilon_x &= \frac{\partial u}{\partial x} - zw_{xx} \quad \varepsilon_y = \frac{\partial v}{\partial y} - zw_{yy} \quad \varepsilon_{xy} = \frac{\partial v}{\partial x} + \frac{\partial u}{\partial y} - 2zw_{xy} \\ \chi_x &= \frac{\partial^2 w}{\partial x^2} \quad \chi_y = \frac{\partial^2 w}{\partial y^2} \quad \chi_{xy} = 2\frac{\partial^2 w}{\partial x \partial y}. \end{aligned} \tag{6.6}$$

It can be shown that the above strain–displacement relation based on shallow shell assumptions is justified when the following relations are satisfied

$$\begin{aligned} &\frac{1}{R_x}\frac{\partial u}{\partial x} + \frac{1}{R_{xy}}\frac{\partial v}{\partial x} \ll \frac{\partial^2 w}{\partial x^2} \\ &\frac{1}{R_y}\frac{\partial v}{\partial y} + \frac{1}{R_{xy}}\frac{\partial u}{\partial y} \ll \frac{\partial^2 w}{\partial y^2} \\ &\frac{1}{R_x}\frac{\partial u}{\partial y} + \frac{1}{R_y}\frac{\partial v}{\partial x} + \frac{1}{R_{xy}}\left(\frac{\partial u}{\partial x} + \frac{\partial v}{\partial y}\right) \ll 2\frac{\partial^2 w}{\partial x \partial y} \end{aligned} \tag{6.7}$$

The problem of a shallow shell was treated as a problem of a plate with initial curvature by Marguerre (1938). In such a case, the shell formulation gets simplified since the displacement components are referred to the Cartesian coordinate directions rather than the shell's middle surface.

Representing the shell surface by a function $w_0(x, y)$, the membrane strains and curvatures for a shallow shell could be written as

$$\begin{aligned} \varepsilon_x &= \frac{\partial u}{\partial x} + \frac{\partial w_0}{\partial x}\frac{\partial w}{\partial x} \\ \varepsilon_y &= \frac{\partial v}{\partial y} + \frac{\partial w_0}{\partial x}\frac{\partial w}{\partial y} \\ \varepsilon_{xy} &= \frac{\partial u}{\partial y} + \frac{\partial v}{\partial x} + \frac{\partial w_0}{\partial x}\frac{\partial w}{\partial y} + \frac{\partial w_0}{\partial y}\frac{\partial w}{\partial x} \\ \chi_x &= \frac{\partial^2 w}{\partial x^2} \\ \chi_y &= \frac{\partial^2 w}{\partial y^2} \\ \chi_{xy} &= 2\frac{\partial^2 w}{\partial x \partial y} \end{aligned} \tag{6.8}$$

When a patch of shallow shell elements approximates the geometry of a general shell, each element behaves like a shallow shell with respect to the local element plane. Nevertheless, the curvature in each element is constant. When the shallow shell elements represent the geometry of a deep shell, the initial slope is discontinuous across the inter-element boundary. Hence, care must be exercised using shallow shell elements to solve deep shell problems.

6.2.3 *Curved Elements Based on Deep Shell Theory*

The use of general free-form shell theory introduces complex algebra, and thus, the improved geometric representation is achieved at the expense of increased effort on shell formulations. The major difficulty in deep shell elements is the satisfaction of inter-element continuity requirements and the representation of rigid body modes. This becomes a crucial issue in the minimum of total potential energy formulation. This prompted many researchers to develop higher-order shell elements, increasing computational effort. As seen from the literature, the effort is still continuing.

6.2.4 *Degenerated Solid Element for Shell Analysis*

The classical shell theory is a special form of the general three-dimensional theory of elasticity. Hence, researchers have made many attempts to begin with a general solid element and then impose constraints on it to transform it into a shell element. This has the advantage of circumventing several restrictions imposed by shell theories. For example, a 20-node 3-D brick isoparametric element is chosen, and a series of constraints is applied to transform it into a shell element. There will be a problem of stiffness in the thickness direction dominating an excessive number of degrees of freedom. Several researchers have addressed these issues in the literature. Nevertheless, among researchers, a debate regarding the 'degenerated solid approach' versus 'shell theories' are reported in the literature (Butcher and Ramm 1992).

6.3 Review of Some Earlier Research Work

A full review of all works on finite element shell instability is extensive, so this chapter highlights key developments in the field. Comprehensive reviews have been provided by Crisfield (1991) and Yang (2000).

One of the earliest significant contributions to the postbuckling theory for shells came from Koiter (1945), who introduced the 'initial postbuckling theory' using perturbation methods to simplify nonlinear equilibrium equations. Koiter (1945) initially focused on small imperfections and later addressed finite imperfections in shells. Early finite element applications to shell problems were carried out by Geis (1977) and Geis and Bergan (1978), who used the co-rotational approach to develop flat shell elements. Reddy (1981) made one of the earliest contributions to the nonlinear analysis of laminated composite shells.

Sheinman and Simitses (1983) developed formulations for imperfect cylindrical shells under axial compression, while Oliver and Onate (1984) used a Total Lagrangian approach for the postbuckling analysis of both plates and shells. Zhang and Mathews (1985) proposed a method for the large deflection behavior of laminated composite shell panels using a modified Galerkin method. Contributions using Marguerre's shell theory were made by Jetteur and Fray (1986), Stolarski et al. (1984), and Fu and Chia (1989), with the latter focusing on the postbuckling behavior of composite panels with shear deformation.

Saigal et al. (1986) introduced curved shell elements for laminated shells, while Chia (1987) presented shallow shell formulations for laminated composite panels on elastic foundations. Yang et al. (1989) introduced a 48-degrees-of-freedom thin shell element based on the Love–Kirchhoff theory, which addressed postbuckling without membrane locking.

Noor and Burton (1990) provided a detailed review of computational models for shell problems, and Librescu and Chang (1992) offered analytical formulations for the postbuckling of multilayered composite curved panels. Palazotto et al. (1992) presented secant matrix formulations for stability analysis of laminated composite shells. Boisse et al. (1994) and Parish (1995) contributed to degenerated shell element formulations for nonlinear analysis, while Onate et al. (1994) introduced a simple triangular shell element free from shear locking, suitable for both thin and thick shells.

Allman (1995) developed a shallow curved shell element with quadratic middle surface geometry to model arbitrary curvatures accurately. Madenci and Barut (1995), Sabir and Djoudi (1995), and Pai and Palazotto (1995) made significant contributions to large deflection analysis of composite shells, while Ganapathy and Varadhan (1995) used the secant matrix technique for nonlinear vibration analysis. Atluri (1997) provided a comprehensive reference for finite element analysis of composite shell postbuckling behavior.

Kim et al. (1998) introduced an Updated Lagrangian formulation for postbuckling analysis of composite shells with initial imperfections. Carrera and Parisch (1998) reviewed geometrically nonlinear multilayered shell models, and Chaudri and Hsia (1999) developed a serendipity-type thick shell element using the Total Lagrangian format in curvilinear coordinates.

Further contributions include Kheyrkhahan and Peek (1999), who developed a general shell element using the Lyapunov-Schmidt-Koiter (LSK) theory, with applications in cylindrical and spherical shell postbuckling. Ferreira and Barbosa (2000) applied Marguerre's shell theory to layered buckling analysis, though the results did not match those of Geis (1977). Teng and Song (2001) addressed issues in modeling shell imperfections and the influence of scaled-down models in postbuckling analysis.

This review highlights significant advancements in shell postbuckling analysis, particularly applying finite element methods and treating shell initial imperfections in both isotropic and composite shells.

Many of the shell forms, which are encountered in practice, are shallow. A literature review shows that many of the general shell formulations presented are too rigorous to be implemented in existing finite element (FE) codes. In almost all the finite element (FE) codes, routines would be available to compute displacement gradients that appear in strain–displacement relations. In the present shell formulation, the secant matrices are functions of displacement gradients; hence, implementing the present formulation in existing FE codes is easy.

6.4 Numerical Formulation for the Postbuckling Analysis of Shallow Shells

Marguerre's shell theory for linear behavior is given in Eq. (6.8). This theory is extended into a large deflection range by adding a square of the flexural strain term as

$$\begin{aligned}\varepsilon_x &= \frac{\partial u}{\partial x} + \frac{1}{2}\left(\frac{\partial w}{\partial x}\right)^2 + \frac{\partial w_0}{\partial x}\frac{\partial w}{\partial x} - z\frac{\partial^2 w}{\partial x^2}\\ \varepsilon_y &= \frac{\partial v}{\partial y} + \frac{1}{2}\left(\frac{\partial w}{\partial y}\right)^2 + \frac{\partial w_0}{\partial x}\frac{\partial w}{\partial y} - z\frac{\partial^2 w}{\partial y^2}\\ \gamma_{xy} &= \frac{\partial u}{\partial y} + \frac{\partial v}{\partial x} + \frac{\partial w}{\partial x}\frac{\partial w}{\partial y} + \frac{\partial w_0}{\partial x}\frac{\partial w}{\partial y} + \frac{\partial w_0}{\partial y}\frac{\partial w}{\partial x} - 2z\frac{\partial^2 w}{\partial x \partial y}.\end{aligned} \tag{6.9}$$

It is seen that the above strain–displacement relation is the same as the ADF formulation explained in Chap. 5. Hence, the studies presented below use the secant matrices ADF framework. Hence, the secant matrix expressions are not repeated here.

6.5 Numerical Studies

This study considers two types of shell postbuckling problems to validate the derived shallow shell formulation. The first category comprises shells loaded out-of-plane, and the second is axially loaded shells.

6.5.1 *Transversely Loaded Shallow Shells*

The first two problems in this category are the standard cylindrical shell panels by Sabir and Lock (1973). The first problem is the simply supported cylindrical shell loaded by a central concentrated load as shown in Fig. 6.4. The thickness of the shell is taken as 6.35 mm. The resulting behavior is a snap-through and snap-back at points A and B, respectively. The results of the present study are compared with the results of displacement at point B with other researchers. It is seen that the present formulation predicts the limit load very accurately. Many researchers (Ramm 1980) have observed that the snap-back portion of the load–displacement curve that Sabir and Lock (1973) presented is a little erroneous.

Regarding the small deviation in the snap-back range, it may be pointed out that such variations exist in the results presented by many authors. There appears to be no accepted benchmark solution. The comparison of the present formulation in the prebuckling range is good and shows a deviation in deep postbuckling ranges. Since Surana(1983) used the displacement control algorithm for path tracing the snap-back behavior is completely missed.

In the second problem, the thickness is increased from 6.35 to 12.7 mm. Since the stiffness of the shell during the postbuckling range increases, the upward movement of the shell in the snap-back portion is reduced. The problem presents snap-through behavior at both ends A and B, as shown in Fig. 6.5. The results are compared with the finite element results of Surana (1983) and the finite strip results of Madasamy and Kalyanaraman (1994). The agreement is good, except for a slight deviation in the deep postbuckling ranges.

The next problem is a composite cylindrical shell shown in Fig. 6.6, with a thickness of 12.7 mm and two layers of ±45° construction. The results of the present study (Fig. 6.6) compare very well with the result of Saigal et al. (1986), who used only 2 × 2 mesh per quarter, and the present study is presented for an 8 × 8 mesh per quarter.

The next problem is that of a clamped shell. The shell ordinates are specified as imperfections for a clamped plate. Various magnitudes of imperfection have been specified and the results are presented in Fig. 6.7. For the case with zero imperfection, the problem reduces to that of a clamped plate. The results of the present study agree very well with the analytical solution of Levy (1942). As the imperfection

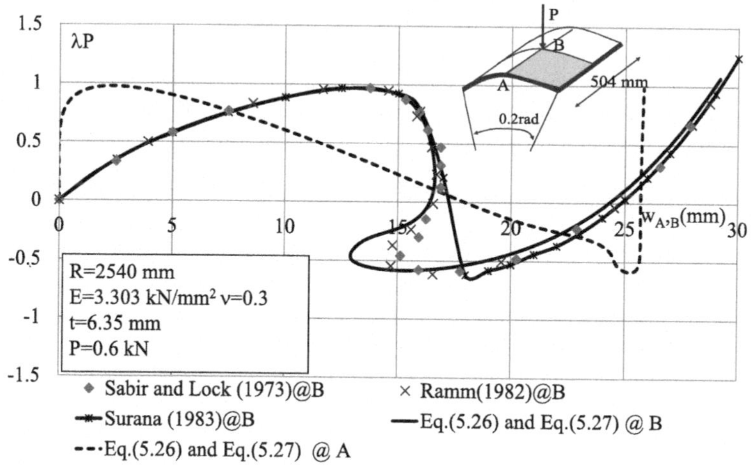

Fig. 6.4 Postbuckling behavior of cylindrical shell subjected to point load (t = 6.35 mm)

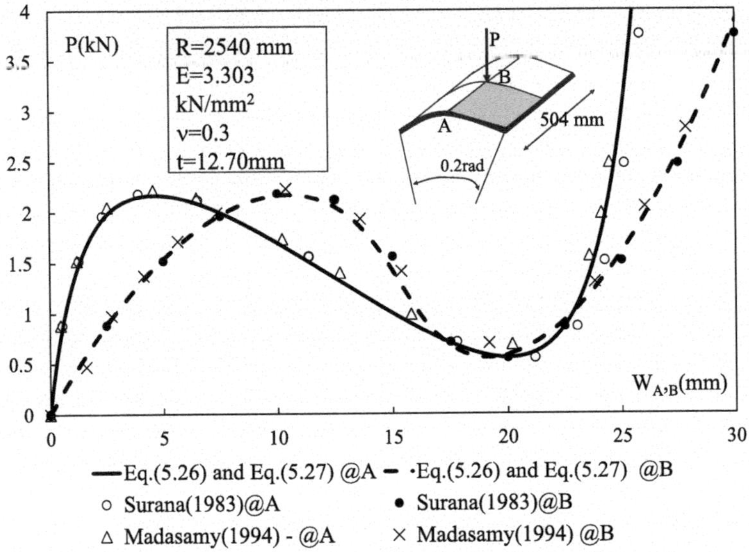

Fig. 6.5 Postbuckling behavior of cylindrical shell subjected to point load (t = 12.7 mm)

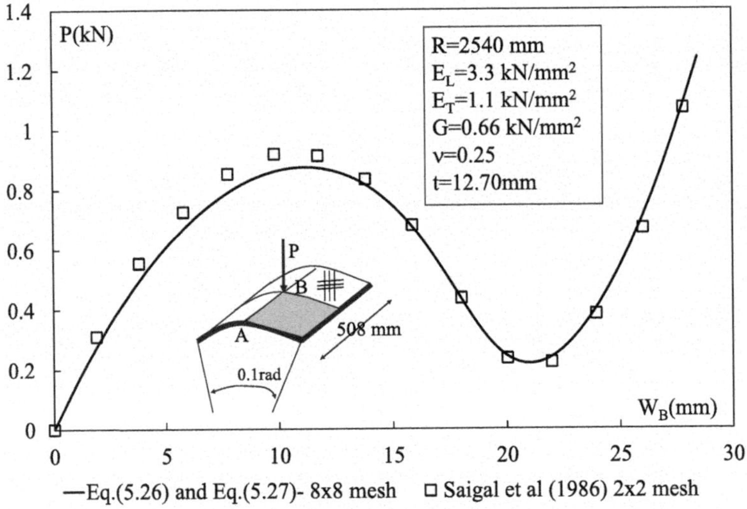

Fig. 6.6 Postbuckling behavior of cylindrical composite shell subjected to point load

shape increases, the behavior gradually changes from stable to limit point behavior. On increasing the shell ordinates by increasing w_0/t, the shell gets bending stiffness because of the arch action. Hence, we see from Fig. 6.7 that as the initial imperfection or shell ordinates are increased, an initial stiffness increase is observed. However, once this initial stiffness is overcome, the shell form unloads at the limit point and

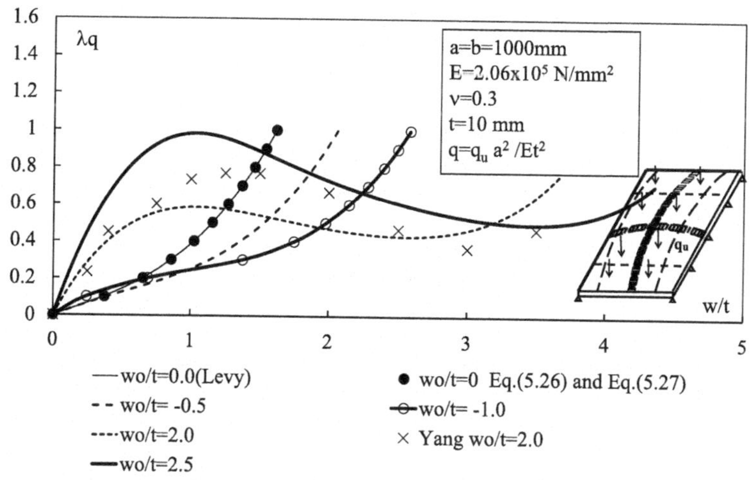

Fig. 6.7 Postbuckling behavior of a clamped shell under UDL

snaps to the other side. It is seen from Fig. 6.7 that the results of Yang's formulation based on total displacements formulation (TDF) $(w + w_0)$ (Yang 1972) for $w_0/t = 2.0$ are much away from the results of the present study for $w_0/t = 2.0$.

The next problem is a spherical shell subjected to a central concentrated load. The details of the spherical shell are shown in Fig. 6.8. The results of the present study are compared with the results of Geis (1977), Noor and Anderson (1981), and Surana (1983) using finite elements. The results of Geis (1977), who used a co-rotational procedure for shell element formulation, are considered a benchmark by many researchers. It is seen that the present study agrees very well with the results of Geis (1977).

6.5.2 Axially Loaded Shallow Shells

In the study of axially loaded shells, the first example chosen is the postbuckling of the shell panel presented by Riks et al. (1990) (Fig. 6.9). The results presented by Riks et al. (1990) assumed that the unloaded edges are immovable. Because of this condition, while the shell panel is axially loaded, the development of uniform axial stress is not realized, and hence, the shell panel does not show bifurcation behavior. The results of the present study agree well with the results of Riks et al. (1990) (Fig. 6.9), except for a slight deviation in the deep postbuckling ranges.

The next problem is that of laminated composite carbon epoxy panels subjected to axial compression (Fig. 6.10). For a 90/0 layup up, postbuckling finite element solutions were presented by Kim et al. (1998). The problem is presented in two parts. In the first case, the shell panel is taken to be perfectly flat and axially loaded.

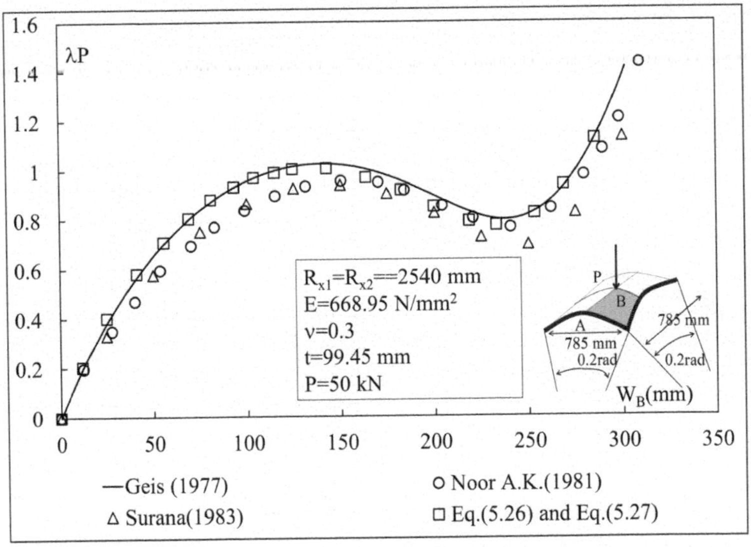

Fig. 6.8 Postbuckling behavior of spherical shell subjected to point load

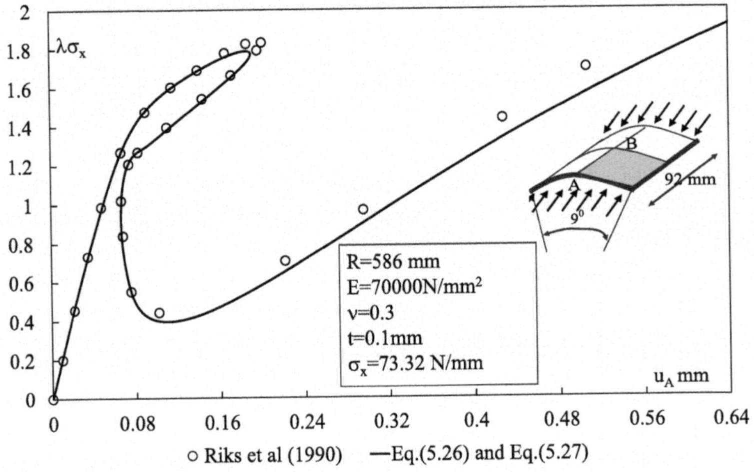

Fig. 6.9 Postbuckling behavior of cylindrical shell subjected to axial compression

Because of the axial flexural coupling of the layup sequence, the plate starts deflecting out-of-plane right from the onset of loading. For such a case, the results of the present study shown in Fig. 6.10 compare very well with the results presented by Kim et al. (1998). The next part is the curved panel with curvature parameter Kr = 25, in which Kr = b^2/rt, where 'b' is the breadth, 'r' is the radius, and 't' is the thickness of the shell panel. The panel exhibits limit point behavior, and the results

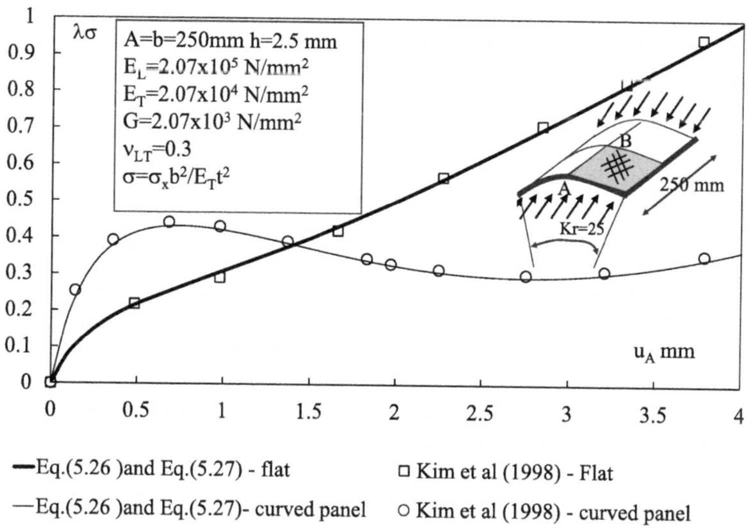

Fig. 6.10 Postbuckling behavior of flat and curved panels subjected to axial compression

of the present study agree very well with the results presented by Kim et al. (1998) (Fig. 6.10).

The next problem taken for comparison is the problem presented by Kroplin (1982). The cylindrical shell panel under compression is shown in Fig. 6.11. Two analyses were considered: one with a local imperfection of 0.1 mm at B and the other without imperfection. It is seen from Fig. 6.11 that the ADF formulations presented in Chap. 5 are quite efficient in modeling the postbuckling behavior of cylindrical shells. This example also exemplifies the capability of the present formulation in modeling localized imperfections.

In order to check the present formulation in modeling shell panels of different curvatures, the results presented by Zhang and Mathews (1985) and Kim (1996) are chosen. It is seen (Fig. 6.12) that the results are in line with Kim's (1996) results.

6.6 Summary

In this chapter, the buckling behavior of shell structures is briefly explained. The different approaches adopted by various researchers, such as the (i) flat facet shell, (ii) shallow shell theories, (iii) deep shell theories, and (iv) degenerated shell theories, have been briefly discussed. The shallow shell theory, which is pertinent to the present study, is dealt with in a little more detail. The development based on Marguerre's shell formulation is presented. Numerical studies comprising isotropic and anisotropic shell panels subjected to both transverse and membrane loads have been carried out, and the results are presented. The present formulation has been demonstrated to be

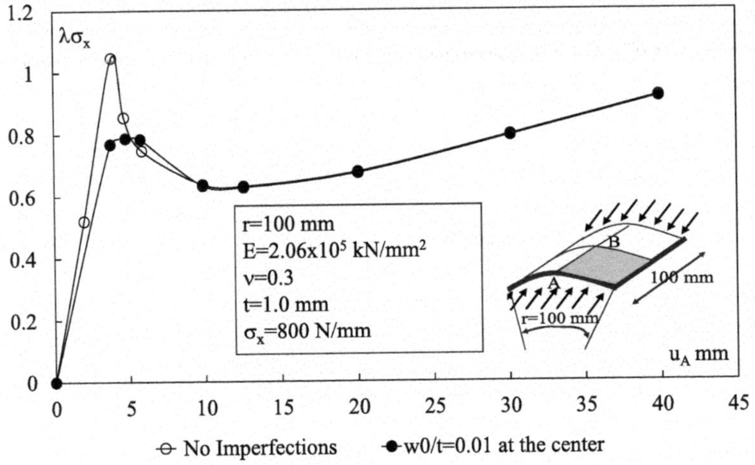

Fig. 6.11 Postbuckling behavior of curved panels with imperfection

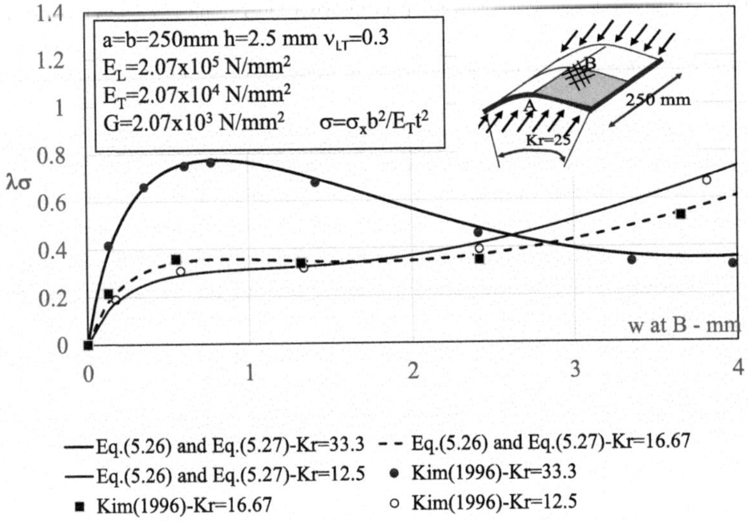

Fig. 6.12 Postbuckling behavior of curved panels with various curvatures

very efficient in the solution of isotropic and composite shell panels subjected to compressive loads.

Appendix

Improved Secant Matrices for the Postbuckling Analysis of Thin, Imperfect, Laminated Composite Plates

The displacement gradient vector '$\mathbf{g}_\alpha$' can be written as

$$\mathbf{g}_a^T = \{u_{,x}, u_{,y}, v_{,x}, v_{,y}, w_{,x}, w_{,y}, w_{,xx}, w_{,yy}, w_{,xy}\}$$

and the comma represents the differentiation of the preceding quantity by the succeeding indices. The nonzero coefficients matrices $K_{\alpha\beta}$, $N0_{\alpha\beta}$, $N1_{\alpha\beta}$, $N2_{\alpha\beta}$ as in Eq. (5.13), are presented below. The coefficients of the matrices A, B, and D could be obtained from Eq. (5.9). The nonzero coefficients of the linear stiffness matrix $\mathbf{K_{ij}} = \mathbf{K_{ji}}$ (9 × 9) matrix are given below

$K(1,1) = A_{11}$	$K(3,3) = K(2,2)$
$K(1,2) = \frac{A_{13}+A_{31}}{2}$	$K(3,4) = K(2,4)$
$K(1,3) = K(1,2)$	$K(3,7) = K(2,7)$
$K(1,4) = \frac{A_{12}+A_{21}}{2}$	$K(3,8) = K(2,8))$
$K(1,7) = B_{11}$	$K(3,9) = K(2,9)$
$K(1,8) = B_{12}$	$K(4,4) = A_{22}$
$K(1,9) = 2B_{13}$	$K(4,7) = B_{21}$
$K(2,2) = A_{33}$	$K(4,8) = B_{22}$
$K(2,3) = K(2,2)$	$K(4,9) = 2B_{23}$
$K(2,4) = \frac{A_{23}+A_{32}}{2}$	$K(7,7) = D_{11}$
$K(2,7) = B_{31}$	$K(7,8) = \frac{D_{12}+D_{21}}{2}$
$K(2,8) = B_{12}$	$K(7,9) = D_{13} + D_{31}$
$K(2,9) = 2B_{33}$	$K(8,8) = D_{22}$
	$K(8,9) = D_{23} + D_{32}$
	$K(9,9) = 4D_{33}$

The nonzero coefficients of the initial imperfection matrix $\mathbf{N0_{ij}} = \mathbf{N0_{ji}}$ (9 × 9) are given below.

$N0(1,5) = A_{11}w_{0,x} + A_{31}w_{0,y} N0(1,6) = A_{21}w_{0,y} + A_{31}w_{0,x}$	$N0(5,5) = A_{11}w_{0,x}^2 + (A_{31} + A_{13})w_{0,x}w_{0,y} + A_{33}w_{0,y}^2$
$N0(2,5) = A_{13}w_{0,x} + A_{33}w_{0,y}$	$N0(5,6) = \frac{A_{21}+A_{12}+2A_{33}}{2} w_{0,x}w_{0,y} + \frac{A_{13}+A_{31}}{2} w_{0,x}^2 + \frac{A_{23}+A_{32}}{2} w_{0,y}^2$
$N0(2,6) = A_{23}w_{0,y} + A_{33}w_{0,x}$	$N0(5,7) = B_{11}w_{0,x} + B_{31}w_{0,y}$
$N0(3,5) = N0(2,5)$	$N0(5,8) = B_{12}w_{0,x} + B_{32}w_{0,y}$
$N0(3,6) = N0(2,6)$	$N0(5,9) = 2B_{13}w_{0,x} + 2B_{33}w_{0,y}$
$N0(4,5) = A_{12}w_{0,x} + A_{32}w_{0,y}$	$N0(6,6) = A_{22}w_{0,y}^2 + (A_{32} + A_{23})w_{0,x}w_{0,y} + A_{33}w_{0,x}^2$
$N0(4,6) = A_{22}w_{0,y} + A_{32}w_{0,x}$	$N0(6,7) = B_{21}w_{0,y} + B_{31}w_{0,x}$
	$N0(6,8) = B_{22}w_{0,y} + B_{32}w_{0,x}$
	$N0(6,9) = 2B_{23}w_{0,y} + B_{32}w_{0,x}$

The nonzero coefficients of first-order secant matrix $\mathbf{N1}_{ij} = \mathbf{N1}_{ji}$ (9 × 9) are given below.

$N1(1,5) = A_{11}w_{,x} + A_{13}w_{,y}$	$N0(5,5) = A_{11}u_{,x} + A_{23}v_{,y} + A_{31}u_{,y} + A_{31}v_{,x} + 6c_1w_{,x} + 2c_3w_{,y} - B_{11}w_{,xx} - B_{12}w_{,yy} - 2B_{13}w_{,xy}$
$N1(1,6) = A_{12}w_{,y} + A_{13}w_{,x}$	$N1(5,6) = A_{13}u_{,x} + A_{23}v_{,y} + A_{31}u_{,y} + A_{33}v_{,x} + 2c_3w_{,x} + 2c_4w_{,y} - B_{31}w_{,xx} - B_{32}w_{,yy} - 2B_{33}w_{,xy}$
$N1(2,5) = A_{13}w_{,x} + A_{33}w_{,y}$	$N1(5,7) = -B_{11}w_{,x} - B_{31}w_{,y}$
$N1(2,6) = A_{32}w_{,y} + A_{33}w_{,x}$	$N1(5,8) = -B_{12}w_{,x} - B_{32}w_{,y}$
$N1(3,5) = A_{31}w_{,x} + A_{33}w_{,y}$	$N1(5,9) = -2B_{13}w_{,x} - 2B_{33}w_{,y}$
$N1(3,6) = A_{32}w_{,y} + A_{33}w_{,x}$	$N1(6,6) = A_{12}u_{,x} + A_{22}v_{,y} + A_{32}u_{,y} + A_{32}v_{,x} + 6c_2w_{,y} + 2c_4w_{,x} - B_{21}w_{,xx} - B_{22}w_{,yy} - 2B_{23}w_{,xy}$
$N1(4,5) = A_{21}w_{,x} + A_{23}w_{,y}$	$N1(6,7) = -B_{21}w_{,y} - B_{31}w_{,x}$
$N1(4,6) = A_{22}w_{,y} + A_{23}w_{,x}$	$N1(6,8) = -B_{22}w_{,y} - B_{32}w_{,x}$
	$N1(6,9) = -2B_{23}w_{,y} - 2B_{33}w_{,x}$

$C_1 = 0.5A_{11}w0_{,x} + 0.5A_{13}w0_{,y}$
$C_2 = 0.5A_{22}w0_{,y} + 0.5A_{23}w0_{,x}$
$C_3 = A_{31}w0_{,x} + 0.5A_{12}w0_{,y} + 0.5A_{13}w0_{,x} + A_{33}w0_{,y}$
$C_3 = A_{21}w0_{,x} + 0.5A_{32}w0_{,y} + 0.5A_{23}w0_{,x} + A_{33}w0_{,x}$

The nonzero coefficients of the second-order secant matrix $\mathbf{N2_{ij}} = \mathbf{N2_{ji}}$ (9 × 9) are given by

$$N2(5,5) = \frac{3}{2}(A_{11}w_{,x}^2) + \frac{3}{2}(A_{13} + A_{31})w_{,x}w_{,y} + \frac{(A_{12} + A_{21} + 4A_{33})}{4}w_{,y}^2$$
$$N2(5,6) = \frac{3}{4}(A_{13} + A_{31})w_{,x}^2 + \frac{3}{4}(A_{23} + A_{32})w_{,y}^2 + \frac{(A_{12} + A_{21} + 4A_{33})}{42}w_{,x}w_{,y}$$

$$N2(6,6) = \frac{3}{2}(A_{22}w_{,y}^2) + \frac{3}{2}(A_{23} + A_{32})w_{,x}w_{,y} + \frac{(A_{12} + A_{21} + 4A_{33})}{4}w_{,x}^2$$

References

Allman DJ (1995) On the assumed displacement fields of a shallow curved shell finite element. Int J Num Meth Eng 11:159–166. https://doi.org/10.1002/cnm.1640110209

Atluri SN (1997) Structural integrity and durability. Tech Science Press, Forsythm, GA

Boisse P, Daniel JL, Gelin JC (1994) A C^0 three node shell element for nonlinear structural analysis. Int J Num Meth Eng 37:2339–2364. https://doi.org/10.1002/nme.1620371402

Butcher B, Ramm E (1992) Nonlinear analysis of shells by finite elements. CISM courses and lectures No. 328, Springer, Wein

Carrera E, Parish H (1998) An evaluation of geometrically nonlinear effects of thin and moderately thick multi-layered composite shells. Compos Struct 40:11–24. https://doi.org/10.1016/S0263-8223(97)00145-1

Chaudri RA, Hsia RL (1999) Effect of thickness on the large elastic deformation behaviour of laminated shells. Compos Struct 44:117–128. https://doi.org/10.1016/S0263-8223(98)00115-9

Chia CY (1987) Nonlinear free vibration and postbuckling of symmetrically laminated imperfect shallow cylindrical panels with mixed boundary conditions resting on elastic foundation. Int J Engg Sci 25:427–441. https://doi.org/10.1016/0020-7225(87)90069-3

Crisfield MA (1991) Nonlinear finite element analysis of solids and structures—vol 1 Essentials. Wiley, New York, NY

Ferreira AJM, Barbosa JT (2000) Buckling behaviour of composite shells. Compos Struct 50:93–98. https://doi.org/10.1016/S0263-8223(00)00090-8

Fu YM, Chia CY (1989) Multimode nonlinear vibration and postbuckling of anti-symmetric imperfect angle-ply cylindrical thick panels. Int J Nonlinear Mech 24:365–381. https://doi.org/10.1016/0020-7462(89)90025-5

Ganapathy M, Varadhan TK (1995) Nonlinear free flexural vibrations of laminated circular cylindrical shells. Compos Struct 30:33–49. https://doi.org/10.1016/0263-8223(94)00025-5

Geis H, Bergan PG (1978) Nonlinear analysis of free-form shells by flat finite elements. Comput Method Appl Mech Eng 16:11–35. https://doi.org/10.1016/0045-7825(78)90030-0

Geis H (1977) Finite element instability analysis of free form shells. Report No. 77/2, Div. of Structural Mechanics, The Norwegian Institute of Technology, The University of Trondheim, Norway

Jetteur P, Fray F (1986) A four node Marguerre element for nonlinear shell analysis. Eng Comput 3:276–282

Kheyrkhahan M, Peek R (1999) Postbuckling analysis and imperfection sensitivity of general shells by the finite element method. Int J Solids and Struct. 36:2641–2681. https://doi.org/10.1016/S0020-7683(98)00129-2

Kim KD (1996) Buckling behaviour of composite panels using the finite element method. Compos Struct 36:3–43. https://doi.org/10.1016/S0263-8223(96)00063-3

Kim KD, Park T, Voyiadjis GZ (1998) Postbuckling analysis of composite panels with imperfection damage. Comput Mech 22:375–387. https://doi.org/10.1007/s004660050369

Koiter WT (1945) Elastic stability and postbuckling behaviour. Doctoral Dissertation at Technische Hooge School, Delft, Netherland also as Technical report TRFFDL-TR-70-25, Wright Patterson Airforce Base

Kroplin B (1982) Postbuckling instability analysis of shells using mixed method. In: Ramm E (ed) Buckling of shells, Univ. of Stuttgart

Levy S (1942) Square plate with clamped edges under normal pressure producing large deflections. NACA Technical Report 740

Librescu L, Chang MY (1992) Imperfection sensitivity and post buckling behaviour of shear deformable composite doubly curved shallow panels. Int J Solids Struct 29:1065–1083. https://doi.org/10.1016/0020-7683(92)90136-H

Madasamy CM, Kalyanaraman V (1994) Analysis of plated structures with rectangular cut-outs and internal supports using the spline finite strip method. Comput Struct 52:277–286. https://doi.org/10.1016/0045-7949(94)90280-1

Madenci E, Barut A (1995) A free formulation based flat shell element for nonlinear analysis of thin composite structures. Int J Numer Methods Eng 37:3825–3842. https://doi.org/10.1002/nme.1620372206

Marguerre K (1938) Zur theric der gekrummfess platte grosser Formanderuny. In: Proceedings of the Fifth international congress of applied mechanics, Cambridge, Massachusetts, pp 93–101

Noor AK, Anderson CM (1981) Mixed models and reduced/selective integration displacement models for nonlinear shell analysis. In: Hughes TGR, Pifko A, Jay A (ed) Nonlinear finite element analysis of plates and shells, ASME, pp 119–146. https://doi.org/10.1002/nme.1620170409

Noor AK, Burton WS (1990) Assessment of computational models for multi-layered composite shells. Appl Mech Rev 43:67–97. https://doi.org/10.1115/1.3119162

Oliver J, Onate E (1984) A total Lagrangian formulation for geometrically nonlinear analysis of structures using finite elements Part I. Two dimensional problems: shell and plate structures. Int J Numer Meth Eng 20:2253–2281. https://doi.org/10.1002/nme.1620201208

Onate EF, Zaarate F, Flores F (1994) A simple triangular element for thick and thin plate and shell analysis. Int J Numer Methods Eng 37:2569–2582. https://doi.org/10.1002/nme.1620371505

Pai PF, Palazotto AN (1995) Nonlinear displacement based finite element analysis of composite shells—a new total Lagrangian formulation. Int J Solids Struct 32:3047–3073. https://doi.org/10.1016/0020-7683(94)00273-Y

Palazotto AN, Chien LS, Taylor WW (1992) Stability characteristics of laminated cylindrical panels under transverse loading. AIAA J 30:1649–1653. https://doi.org/10.2514/3.11113

Parish H (1995) A continuum-based shell theory for nonlinear applications. Int J Numer Methods Eng 38:1855–1883. https://doi.org/10.1002/nme.1620381105

Ramm E (1980) Strategies for tracing the nonlinear response near limit points. In: Wunderlich W (ed) Nonlinear finite element analysis in Structural Mechanics. Springer Verlaag, Berlin. https://doi.org/10.1007/978-3-642-81589-8_5

Reddy JN (1981) A finite element analysis of large deflection bending of laminated anisotrophic shells. In: Hughes TJR, Pifco A, Jay A (eds) Nonlinear finite element analysis of plates and shells. ASME, pp 249–259

Riks E, Brogan FA, Rankin CC (1990) Numerical aspects of shell instability analysis. In: Kratzig WB, Onate E (eds) Computational methods for non-linear response of shells. Springer Verlag, pp 125–151

Sabir AB, Lock (1973) The application of finite elements to the large deflection geometrically nonlinear behaviour of cylindrical shells. In: Brebbia CA, Tottenham H (eds) Variational Methods in Engg. Southampton University Press, pp 7176–7175

Sabir AB, Djoudi MS (1995) Shallow shell finite element for the large deflection geometrically nonlinear analysis of shells and plates. Thin Walled Struct 21:253–267. https://doi.org/10.1016/0263-8231(94)00005-K

Saigal S, Kapania RK, Yang TY (1986) Geometrically nonlinear finite element analysis of imperfect laminated shells. J Compos Mater 20:197–213. https://doi.org/10.1177/002199838602000206

Sheinman I, Simitses GJ (1983) Buckling and post buckling of imperfect cylindrical shells under axial compression. Comput Struct 17:471–481. https://doi.org/10.1016/0045-7949(83)90016-0

Stolarski H, Belytschko T, Carpenter N, Kennedy JM (1984) A simple triangular curved shell element for collapse analysis. Engg Comp 1:210–218. https://doi.org/10.1108/eb023574

Surana KS (1983) Geometrically nonlinear formulation for the curved shell elements. Int J Numer Methods Eng 19:581–615. https://doi.org/10.1002/nme.1620190409

Teng JG, Song CY (2001) Numerical models for nonlinear analysis of elastic shells with eigen-mode affine imperfections. Int J Solids Struct 38:3263–3280. https://doi.org/10.1016/S0020-7683(00)00222-5

Vlasov VZ (1957) Cylindrical shells and new ways for developing thin walled spatial systems in structural mechanics. Proceedings of the second symposium on concrete shell roof construction, Oslo 121–146

Yang TY (1972) Elastic snap through analysis of curved plates using discrete elements. AIAA J 10(4):371–372

Yang TY (2000) A survey of recent shell finite elements. Int J Num Methods Eng 47:101–127. https://doi.org/10.1002/(SICI)1097-0207(20000110/30)47:1/3%3c101::AID-NME763%3e3.0.CO;2-C

Yang TY Kapania RK Saigal S (1989) Accurate rigid body mode representation for a nonlinear curved thin-shell element. AIAA J 27(2):211–218

Zhang Y, Mathews FL (1985) Large deflection behavior of simply supported laminated panels under inplane loading. J Appl Mech ASME 52:553–585. https://doi.org/10.1115/1.3169100

Chapter 7
Closing Remarks

Abstract This chapter summarizes the reviews, insights, and methodologies explored throughout the book, focusing on advanced computational techniques for postbuckling analysis of structural systems. It overviews core concepts from earlier chapters, such as fundamental stability theories, including critical points and bifurcation, and discusses implementing the secant matrix technique based on the total potential energy (TPE) framework. The chapter emphasizes the practical application of the secant matrix technique in analyzing the postbuckling behavior of various structural forms, trusses, beams, frames, thin-walled sections, plates, and shells. Iterative and incremental computational methods, including advanced path-following methods like the arc-length approach, are highlighted. This book offers a comprehensive framework to analyze postbuckling phenomena in structural engineering practice.

Keywords Variable order secant matrix technique · Postbuckling analysis · Computational stability

7.1 Overview of the Book

The book has covered various geometric nonlinear finite element analysis topics, focusing primarily on postbuckling behavior in structural systems like space trusses, beams, frames, plates, and shells. The book's primary purpose is to demonstrate the application of the **secant matrix technique**—a computational method well-suited for handling large deformations and the postbuckling behavior of structural systems.

Throughout the book, the derived formulations use the secant matrix approach in the **Total Lagrangian (TL) format**, except in the case of trusses where the **co-rotated TL format** is used. In contrast, a Variable Order Secant Matrix (VOSM) technique was developed for beams and frames where large rotations and deformations are expected. The formulation was extended to plates and shells, incorporating initial imperfections for the postbuckling analysis. These formulations are implemented in

A. Jayachandran, *Elastic Postbuckling Analysis of Structural Systems*, Solid Mechanics and Its Applications 232, https://doi.org/10.1007/978-3-031-89936-2_7

a custom computational tool that provides a solution methodology capable of tracing complex postbuckling paths, including **snap-through** and **snap-back** behavior.

7.2 Secant Matrix Technique Applied to Different Structural Systems

The foundation of this book lies in applying energy principles and finite element methods for analyzing structural stability and postbuckling behavior. The secant matrix technique is amenable because it can be adapted to different structural systems and loading conditions.

The **secant matrix technique** is based on expanding the total potential energy of a structure expressed in a **series form**, making it possible to approximate complex nonlinear behavior by using higher-order terms in the expansion. The use of secant matrices allows us to capture the gradual changes in stiffness that occur as a structure deforms.

For **space trusses**, we introduced a co-rotated formulation that is easy to handle the rigid body motions that often dominate the postbuckled state.

For **beams and frames**, the basis is different. Here, large rotations dominate the behavior in the postbuckled state, which requires the development of the Variable Order Secant Matrix (**VOSM) technique**. The order of the secant matrices can be varied based on the degree of nonlinearity in the structure, making it flexible and capable of capturing the large deformations observed in real-world structures.

7.3 Trusses, Beams, Frames, and Thin-Walled Open Sections: Insights from Nonlinear Behavior

Trusses, beams, frames, and thin-walled open sections present unique challenges in structural analysis due to their susceptibility to large deformations and rotations.

In the case of **trusses**, the co-rotational formulation allows for an effective modeling of member instabilities. The technique used in this book builds on a simple geometric principle: the relationship between the load and displacement of the structure in space. This CR-TL formulation has been tested against benchmark problems, published in the literature. For students and researchers, the benefit of this approach lies in its efficiency in focusing on the rigid body geometric changes and member forces separately.

The Variable Order Secant Matrix (VOSM) technique provides an elegant solution to the complex problems in beams and frames. These systems are known for undergoing large rotations and displacements, particularly when subjected to high lateral and gravity forces. Traditional linear methods fail in these cases, leading to inaccurate predictions. The VOSM approach ensures that the predictions remain reliable even in extreme cases—such as a straight cantilever beam being bent into a full circle. Although this is a problem of extreme geometry change, the book demonstrates significant improvements in accuracy without requiring complex calculations.

7.4 Plates and Shells: The Role of Imperfections

One of the most interesting aspects of postbuckling analysis is the role of initial imperfections, particularly in **plates** and **shells**. These imperfections, which often occur due to manufacturing tolerances or material defects, can drastically affect a structure's stability.

For **plates**, especially thin laminated composite or isotropic plates, the secant matrix technique is extended to handle **initial imperfections**. Traditional methods often fail to accurately model plates with imperfections larger than 0.2 times the plate thickness. However, the improved secant matrices developed in this work extend the range of applicability to imperfections as large as the total plate thickness. This advancement is critical for real-world applications where imperfections of this magnitude are expected.

The secant matrix technique is applied in shallow shells using **Marguerre's shell theory**. By treating the shell's curvature as an initial imperfection, this method accurately models the postbuckling behavior of shallow shell structures. Shells are more complex than plates because of their curvature and sensitivity to initial imperfection, which introduces additional challenges in modeling.

7.5 Computational Methods and Path-Tracing Algorithms

Another component of this book has been the emphasis on **numerical methods** for solving nonlinear problems. Several solution techniques have been explored and applied, including the Newton–Raphson, arc-length, and minimum residual displacement (MRD) methods.

One of the features of path tracing in the postbuckled state is the use of the **current stiffness parameter** as an indicator for structural stability. This parameter, updated as the structure deforms, helps guide the solution process, automatically adjusting load increments based on the severity of nonlinearity in the problem. This indicator makes it possible to track complex behaviors, such as **snap-through** or **snap-back** phenomena, without the solution process becoming unstable.

While applicable in many situations, the arc-length method was insufficient for cases with sharp snap-back behavior. The **MRD method,** introduced by Prof. SL Chan, ensures that the path-tracing algorithm remains robust even in highly nonlinear situations.

7.6 Conclusion

This book has introduced a range of secant matrix methods in the nonlinear postbuckling analysis of structural systems. The focus has been on applying the **secant matrix technique** to various structural systems, each presenting its unique challenges.

The co-rotated formulations for trusses, the VOSM technique for beams and frames, and the improved secant matrices for initially imperfect plates and shells all represent valuable contributions to computational structural stability. Also, the techniques offer a robust framework for tackling structural systems' complex, nonlinear behavior under extreme loads.

As you progress in your study of structural stability, I hope the methods presented in this book serve as a foundation for further exploration into the fascinating world of postbuckling behavior of structural systems.

The manufacturer's authorised representative in the EU is Springer Nature Customer Service Centre GmbH, Europaplatz 3, 69115 Heidelberg, Germany. If you have any concerns regarding our products, please contact ProductSafety@springernature.com

Printed and bound by CPI Group (UK) Ltd, Croydon, CR0 4YY
05/10/2026
02228157-0002